THE
NEXT STEP
Finding and Viewing
MESSIER'S OBJECTS

Ken Graun

ken

A Ken Press Book

"Bringing Astronomy to Everyone"

Published by Ken Press, Tucson, Arizona USA
(520) 743-3200 or office@kenpress.com

Other Ken Press books by Ken Graun

Touring the Universe: A Practical Guide to Exploring the Cosmos thru 2017
What's Out Tonight? 50 Year Astronomy Field Guide, 2000 to 2050
Our Earth and the Solar System
Our Galaxy and the Universe
Our Constellations and their Stars
David H. Levy's Guide to the Stars (planispheres, co-author)

Visit
www.whatsouttonight.com

Publisher's Cataloguing-in-Publication Information

Graun, Ken.
    The next step : finding and viewing Messier's objects / by Ken
Graun. – Tucson, Ariz. : Ken Press, 2005.
      p. ; cm.
    Contains biographical information about Charles Messier,
a complete set of star charts, and 150 photographs.
    Includes bibliographical references and index.
      ISBN: 1-928771-12-2
      1. Stars–Clusters. 2. Nebulae. 3. Galaxies. 4. Astronomy–Amateurs' manuals.
5. Astronomy–Observers' manuals. 6. Astronomy–Charts, diagrams, etc.
7. Messier, Charles. I. Title. II. Finding and viewing Messier's objects.

QB851.G73 2005   2004095612
523.8/5–dc22   0505

— PRINTED IN HONG KONG BY SOUTH SEA INTERNATIONAL PRESS LTD —

Table of Contents

Introduction

I wrote this book for the budding amateur astronomer who wants to forge ahead, to go beyond observing the Moon, Planets and Sun. It is a guide that takes one farther into the depths of space, providing an introduction to a body of heavenly wonders commonly called Deep Sky Objects: star clusters, nebulae and galaxies.

During the 1700s, Charles Messier compiled the very first catalogue of deep sky objects from a small Paris observatory. Although Messier was an early pioneer in the field of astronomy, his catalogue of just over 100 objects has not, like many other works, been forgotten with the passage of time.

The allure of his catalogue remains to this day because its objects are the crème de la crème — the biggest and brightest — the most splendid in the heavens. And, what is even better, observing them is well within the grasp of anyone wanting to do so, because all it takes is the smallest modern telescope. There is one other notable point about his catalogue: It contains at least one example from every major category of celestial objects. So if you endeavor to observe this litany, you will experience a journey into the very depths and nature of the Universe itself.

A goal I had in writing this book was to make it as user-friendly as possible. To this end, I sectioned the historical chapter into parts so that its topics would be more accessible and interesting to read. For the actual catalogue, I gave each Messier object a standardized spread to facilitate reference and comparison. You will also find that I repeat information often, solely to reduce the amount of page turning. In short, I did my best to make this an easy-to-use guide so you can keep focused on what's important — finding and viewing these wonderful deep sky denizens.

May your journey in discovery of the Messier objects be just as enjoyable and rewarding as my research and writing of this book have been.

Clear Skies!

K Clgram

Ken Graun
January 2005
Tucson, Arizona

Acknowledgements

I want to wholeheartedly thank everyone who assisted in making this book become more than what I envisioned:

Isabelle Houthakker translated Messier's original descriptions from French into English and proofread the final manuscript.

Larry Moore carefully checked the catalogue section against references for accuracy.

Professor Owen Gingerich at the Harvard-Smithsonian Center for Astrophysics graciously provided information and images on historical documents related to Messier.

Gina Westbeld helped with the history of Paris during the 1700s, but more importantly, she was able to find a copy of Jean-Paul Philbert's book on Messier, while in France.

Paola Bortolotti-van Loon translated portions of Philbert's book which provided new insights into Messier's early years and sojourn to Paris.

Richard Gliech arranged my research visits in Paris and afterwards translated several historical documents.

In Paris, curator Julia Fritsch at the Musée National du Moyen Age was gracious enough to share time and documents adding to Messier's history. Also, the staffs at l'Observatoire de Paris and the Académie des Sciences Archives were most helpful in providing period pictures and allowing me to make copies of historical documents.

While at the Paris Observatory, I was fortunate enough to meet Ulrich Paessler, working on his doctoral dissertation, who kindly interpreted for me.

I want to especially thank Monsieur Roger Leboube from Senones, France for inviting me into his home, the very same house that Messier's oldest brother Hyacinthe lived in.

Finally, I need to acknowledge the help of my assistant, Carolyn Randall, and the love of my wife, daughter and parents.

References/Photographs/Credits

In this book, I am light on footnotes; however, if anyone is interested in a particular citing or the photographs, please contact me at ken@kenpress.com.

Butler, John. *John Louis Emil Dreyer*. Retrieved July 6, 2004 from Armagh Observatory Web site: http://www.arm.ac.uk/history/dreyer.html.

Gay, Peter. *The Enlightenment: An Interpretation*. New York: Norton, 1977.

Jardine, Lisa. *Ingenious Pursuits*. New York: Anchor Books, 2000.

Mallas, John H., and Evered Kreimer. *The Messier Album*. Cambridge: Sky Publishing Corp., 1978.

L'Observatoire de Paris. *Pierre-François-André Méchain*. Retrieved March 16, 2004 from Web site: http://www.obspm.fr/histoire/acteurs/mechain.fr.shtml.

O'Connor, J. J. and E. F. Robertson. *Joseph-Jérôme Lefrançais de Lalande*. Retrieved August 2, 2004 from University of St. Andrews, Scotland: The MacTutor History of Mathematics Web site: http://www-history.mcs.st-andrews.ac.uk/history/Mathematicians/Lalande.html.

O'Meara, Stephen J. *Deep-Sky Companions: The Messier Objects*. Cambridge: Sky Publishing Corp., 1998.

Philbert, Jean-Paul. *Le furet des comètes*. Sarreguemines, France, 2000.

Sinnot, Roger W., ed. *NGC 2000.0*. Cambridge: Sky Publishing Corp., 1988.

Students for the Exploration and Development of Space (SEDS), The Messier Catalog. Web site: http://www.seds.org/messier/.

Toomer, G. L. *Ptolemy's Almagest*. Princeton: Princeton University Press, 1998.

Yeomans, Donald K. *Comets — A Chronological History of Observation, Science, Myth, and Folklore*. New York: John Wiley & Sons, 1991.

Throughout this book, photos not credited are by Ken Graun.

Cover designed by Debra Niwa. Cover photo credits: M1 Crab Nebula: Jay Gallagher (U. Wisconsin), N.A. Sharp (NOAO)/WIYN/NOAO/NSF. M22 globular cluster: N.A. Sharp, REU program/NOAO/AURA/NSF. M31 Andromeda Galaxy: Bill Schoening, Vanessa Harvey/REU Program/NOAO/AURA/NSF. M42 Orion Nebula: Bill Schoening/NOAO/AURA/NSF. M45 Pleiades: Anglo-Australian Observatory/Royal Observatory Edinburgh. M57 Ring Nebula: NASA and the Hubble Heritage Team (STSCI/AURA). Pages 4 & 6 show a reverse image of a "mark" Messier drew on a personal copy of his final catalogue, courtesy of Owen Gingerich.

Charles Messier (1730–1817)
The Man, His Friends and the Times

Who was Charles Messier? Simply put, he was a small-town lad from a good family who acquired a job in the big city, and because of hard and diligent work, became an accomplished observational astronomer and a famous discoverer of comets.

However, among amateur astronomers, Messier is best known as the cataloguer of 103 (his final published catalogue count) of the biggest and brightest deep sky objects, that is star clusters, nebulae and galaxies, which can be seen from the northern hemisphere. What makes these objects special is that all of them are visible with the smallest of modern telescopes, so they often become the first set of deep sky objects observed by amateurs.

To understand someone, you cannot divorce them from the era in which they lived. Messier's life spanned most of the 1700s, which is often referred to as the "Age of Enlightenment." It was a time in Europe when scholarly thinking, scientific exploration and literary license were encouraged and acted upon. However, it was also a time of civil unrest with social and political reforms spawning revolutions and wars.

The "story" of Messier's catalogue is actually the work of two people: Messier and Pierre Méchain. Messier catalogued 62 objects before meeting Méchain; thereafter, both contributed to its entries. In regards to this historical catalogue, Méchain's role is often overlooked, which is unfortunate since he was a major contributor and more importantly, Messier's friend.

The house where Charles Messier was born and raised still stands in Badonviller, France. Since his father held a position akin to that of mayor, it was one of the nicest in town. In honor of Charles, a medallion was placed on its façade in 1930 (positioned to the left of center, between the floors of shuttered windows — see enlargement on page 15).

Reader's Note

For those with a particular interest in the history of Messier, I have interwoven the chronology and details of his life throughout the various parts and sidebars of this chapter.

Messier's official portrait, a pastel painted in 1771 by Ansiaume on the occasion of his appointment as Astronomer of the Navy. The red-ribboned medal stands for the French Academy of Sciences. On the back of this painting he wrote, *"This portrait is a good likeness except that I appear younger than I am and I have been given a better expression than I have."*

Photo courtesy of the Observatoire de Paris.

Charles Messier
Observational Astronomer and Comet Hunter
Born in Badonviller, France on June 26, 1730.
Passed away in Paris, France on April 12, 1817.

Dubbed by King Louis XV, *"The Comet Ferret."*
Member of the French Academy of Sciences.
Held the position of Astronomer of the Navy.
Member of the Bureau of Longitudes.
Napoleon bestowed on him the *Legion of Honor* medal.

Early years

Charles Messier was born in the walled fortress town of Badonviller (pronounced Bah Dohn Villay). Soldiers guarded it until the late 1600s and its two drawbridges provided access as well as protection.

Badonviller is located about 200 miles east of Paris, southeast of the city of Nancy, which places it 40 miles west of the German border. The area is replete with forests and the town straddles a hill set within beautiful rolling countryside shaped by the gently sloping Vosges mountain range. Mount Donon, its peak, sits just a dozen miles east.

Badonviller is in the French province of Lorraine, located in the region of Salm. Lorraine and Salm are family names of dukes, counts, princes, etc., who held protectorate control over their domains. Needless to say, territorial strife existed between families, but was somewhat allayed after the marriage of Princess Christine de Salm to Count François II de Lorraine in 1598 (it was not until the French revolution of 1796 that serfdom ended). With this marriage of convenience, control over some towns was divided and in some instances,

The Players' Pronunciation Guide

Messier (Mess-ee-ay). First published cataloguer of deep sky objects.
Méchain (Mih-Shan). Colleague and friend of Messier who helped with deep sky cataloguing. He also searched and found comets.
Delisle (Da-Leel). Messier's employer and astronomy mentor.
Lalande (La-Lond). "Colorful" French astronomer who was a colleague and friend of Delisle, Méchain and Messier.

Below. A panoramic view of the countryside on the imme-diate outskirts of Badonviller, where Charles Messier grew up. The 130-foot church bell-tower that sits atop the town's hill can be seen jutting above the horizon (to the right of the road and above the tilled field). Although this area is flattish, steep hills rise up just a mile away.

individual houses within towns were split between controlling parties, but Badonviller became the rural "capital" for the administration until a major reshuffling in 1751.

By the late 1500s, Badonviller and the surrounding area was known for its quality pistol production, which was supplanted in the 1600s by tanneries that produced some of the finest European leather. However, starting in 1724, and on to the present, this region has been known for its production of faience: glazed, color-coated pottery and porcelain.

Nicolas Messier (1682–1741), Charles' father, was also born in Badonviller.

St. Martin's church in Badonviller is about 500 feet from Messier's house. It sits near the hilltop, rising above the rest of the town, and has bells that chime every quarter hour, with an extended musical flourish on the hour.

He married Françoise Grandblaise (died 1765) in her hometown of Senones (8 miles south, through the forest and steep hills, from Badonviller) on January 21, 1716. They had 12 children, but only six survived into adulthood. Charles was the tenth child, with three surviving older brothers, Hyacinthe (1717–1791), Claude, Nicolas-François and a younger brother Joseph and sister Barbe. Hyacinthe was 13 years older and Barbe 3½ years younger than Charles.

Charles was born and baptized on June 26, 1730 in an upscale house that was centrally located in town. His family was financially secure because his father was an

Messier & Méchain: Timeline with Pertinent History

1679 First annual publication of *Connaissance des Temps* (translated as "Knowledge of the Times," an astronomical almanac), which later published Messier's final catalogue. This is still published annually to this day.

1687 Newton publishes his theory of gravity.

1688 Joseph Delisle, Messier's future boss, is born.

1699 French Royal Academy of Sciences established by Louis XIV.

1715 Louis XIV dies — Louis XV becomes king of France.

1727 Newton dies.

1730 **Charles Messier is born on June 26 in the small town of Badonviller, France, 200 miles east of Paris.**

1732 Joseph Lalande is born.

1744 **Pierre Méchain is born on August 16 at Laon, located 75 miles northeast of Paris.**

1751 **On September 23, Messier, at the age of 21, leaves his hometown of Badonviller for Paris to work under astronomer Joseph Delisle.**

1753 **Messier observes and documents Mercury's transit of the Sun, an early milestone in his career.**

1757 **Messier starts looking for Comet Halley.**

1758 **In August, at the age of 28, Messier discovers his first comet at the head of Taurus and also comes across M1 which looks so much like a comet that he notes its position and contemplates a catalogue of comet-like looking objects.**

1759 **Messier independently finds Comet Halley in late January below Pisces.**

1764 **Messier is made a foreign member of the Royal Society of London.**

1765 Delisle retires as Astronomer of the Navy.

1768 Delisle dies.

1769 Napoleon Bonaparte is born.

1770 **Messier marries Marie Vermanchampt, age 37. Messier is finally elected to the French Royal Academy of Sciences.**

1771 **Messier becomes Astronomer of the Navy. Official portrait is painted by Ansiaume. First version of Messier's catalogue, up to M45, is prepared for publication in the 1772 *Mémoires de l'Académie des Sciences*.**

1772 **Both Messier's wife and newborn son die within 11 days of his birth.**

1774 **Messier meets Méchain. Méchain is appointed a Calculator or Surveyor of the Navy — his first assignment involves surveys of French coastlines.** Louis XVI becomes king of France.

1776 Benjamin Franklin meets with the French for support of war with Britain.

1778 France declares war on Britain and sends troops to North America.

1780 **Second edition of the catalogue up to M68, with collaboration by Méchain, printed in the 1783 edition of the *Connaissance des Temps.***

1781 **Messier and Méchain's final version of the catalogue, up to object 103, is printed in the 1784 edition of the *Connaissance des Temps.* In November, Messier is severely injured when he falls into a 25-foot deep ice cellar. He does not fully recover until November, 1782. Méchain discovers his first two comets.** William Herschel in England discovers Uranus.

1782 William Herschel begins his deep sky survey inspired by Messier's catalogue.

1783 Treaty of Versailles — France recognizes independence of the U.S.

1785 Napoleon becomes second lieutenant at the age of 16. Construction begins on Dunsink Observatory in Ireland where Dreyer will work.

1787 U.S. Constitution approved.

1789 French revolution begins with the storming of the Bastille fortress in Paris on July 14, and continues to the middle of 1794. The existing government remains in a state of chaos with **Messier losing his salary**.

1790 Construction of Armagh Observatory begins in Ireland, where Dreyer will become a director.

1791 Metric system adopted in France. Louis XVI accepts a constitution with a legislative assembly formed.

1792 France goes to war with Austria. Parisian mob establishes a new mayor. Mobs kill royalist sympathizers across the country. Monarchy is abolished and Louis XVI is "convicted" of treason.

1793 Royal Academy is disbanded. "Year of Terror" in France with Louis XVI and others dying at the guillotine in Paris.

1795 **Méchain is elected as a member to the new Bureau of Longitudes.** The French capture Amsterdam. At age 26, Napoleon rises to the rank of general and offers a prize for a practical means of preserving food. New academy of sciences is established.

1796 **Messier is elected as a member of the Bureau of Longitudes after Cassini IV resigns.**

1797 Napoleon has a string of victories across Europe.

1798 Laplace of France predicts the existence of black holes. Napoleon heads a French expeditionary force into Egypt.

1799 **Méchain discovers his last two comets.** Napoleon rides to Paris and takes control of the French government. French soldiers discover the Rosetta Stone in Egypt.

1800 **Méchain is appointed director/manager of the Paris Observatory.**

1801 **Messier discovers his last comet on July 12.** Piazzi of Italy discovers the first asteroid, Ceres.

1804 **Méchain, at the age of 60, dies September 20, 1804 at Castillion de la Plana, Spain from yellow fever.** Napoleon names himself emperor of France.

1806 **Napoleon presents Messier with the Cross of the Legion of Honor.**

1807 Lalande dies at age 75.

1809 College baccalaureate examination established in France.

1814 Napoleon abdicates and is exiled.

1815 **Messier suffers a stroke and is partially paralyzed.** Napoleon returns and is defeated at the battle of Waterloo. Napoleon is deported to Santa Helena, an island 1,000 miles west of lower Angola, Africa.

1817 **At the age of 87, Charles Messier passes away on the night of April 11–12 at his residence, Hotel Cluny in Paris.**

1821 Napoleon dies of a stomach ulcer.

1839 First observatory in U.S. founded at Harvard with a 15-inch refractor.

1845 72-inch diameter telescope, later used by Dreyer, is completed at Birr Castle (in the town of Birr), Ireland.

1846 Galle, in Germany, observes Neptune from calculations by Le Verrier of France.

1852 J.L.E. Dreyer is born.

1874 Dreyer becomes an assistant at Birr Castle.

1878 Dreyer becomes an assistant at Dunsink Observatory in Ireland.

1882 Dreyer becomes the fourth director of Armagh Observatory in Ireland where he compiles the NGC & IC catalogues.

1926 J.L.E. Dreyer passes away at Oxford, England on September 14 at the age of 74.

Cataloguing and Comets of Messier & Méchain

1758 Messier, at the age of 28, discovers his 1st comet. **He also notes the position of M1 because it resembles a comet.**

1759 Messier discovers his 2nd comet, the returning Comet Halley.

1760 **Messier records M2** and discovers his 3rd and 4th comets.

1763 Messier discovers his 5th comet.

1764 **Messier records M3 to M40** and discovers his 6th comet.

1765 **Messier records M41.**

1766 Messier discovers his 7th and 8th comets.

1769 **Messier records M42 to M45** and discovers his 9th comet.

1770 Messier discovers his 10th and 11th comets.

1771 **First catalogue preparation up to M45** for publication in the 1772 volume of *Mémoires de l'Académie des Sciences.* **Also in this year, he records M46 to M49 and notes M62 but does not determine its position until 1779.** Messier discovers his 12th comet.

1772 **Messier records M50.**

1773 **Messier observes and notes M110,** draws it alongside the Andromeda Galaxy, in 1807 but does not include it in any catalogue. Messier discovers his 13th comet.

1774 **Messier notes M51 and M52.** Messier meets Méchain.

1777 **Messier records M53.**

1778 **Messier records M54 and M55.**

1779 **Messier records M56 to M62** and discovers his 14th comet. **Méchain discovers/records M63.**

1780 **Messier and Méchain record objects M64 to M79. Second catalogue prepared up to object M68** for publication in the 1783 edition of *Connaissance des Temps.* Messier discovers his 15th comet.

1781 **Messier and Méchain record objects M80 to M100 with Méchain hastily adding M101 to M103. Final catalogue of 103 objects prepared** for the 1784 edition of *Connaissance des Temps.* **After submission, Messier** and Méchain discover several other objects that are never published as entries (M104, M105, M106, M108, and M109). Méchain at the age of 37 discovers his first two comets.

1782 **Méchain records M107, the last formally observed and noted object.** Cataloguing comes to a end primarily because William Herschel begins his cataloguing, listing 1,000 objects by 1785 and 2,500 by 1802.

1785 Messier and Méchain co-discover a comet, their 16th and 3rd respectively.

1786 Méchain discovers his 4th comet.

1787 Méchain discovers his 5th comet.

1788 Messier discovers his 17th comet.

1790 Méchain discovers his 6th comet.

1793 Messier discovers his 18th comet.

1798 Messier discovers his 19th comet.

1799 Méchain discovers his 7th and 8th comets, his last.

1801 Messier discovers his 20th comet, his last.

administrator, holding a mayoral-type position in the community. Nicolas executed duties for the Princes of Salm, acting as a tax collector, a confiscation commissioner who oversaw seized properties, and a judge, settling misdemeanors. He was well respected and performed these duties for a good number of years. Nicolas' family members, friends and acquaintances held similar or other positions in the surrounding communities, so it is not surprising that Charles' godparents were also from the families of a mayor and ex-mayor.

Nicolas was able to arrange for his eldest son, Hyacinthe, to receive training with a procurator, an official entrusted with the management of governmental financial affairs, in the city of Nancy. Hyacinthe completed his training and returned to Badonviller in 1740 to work as an adjudicator, and was responsible for auction sales. Barely a year later, in 1741, when Charles was just 11 years old, tragedy struck the Messier family — Nicolas, their father, died. The family held together, with Hyacinthe becoming head of the household. Within a year, after Charles turned 12, Hyacinthe made him his personal clerk, providing training and obviously giving him brotherly support.

Charles grew up in what can only be described as an idyllic setting. The locale, at the time of his youth, was peaceful and safe. For most kids, this countryside would have been a dream playground because it was abundant with forests, hills, rivers and even a lake. Philbert reports that as a youth, Charles was "courageous, independent, not particularly unruly and a little daredevilish." This, at one particular point, resulted in him falling out of a window and breaking his leg. He completely recovered and was looked after by a benevolent man in the community.

Charles experienced nature firsthand since he lived in the countryside, so his curiosity about the world must have been kindled at an early age. In this setting, astronomy was never far from anyone's mind because the rhythm of rural life, especially farming was based on the cycles of the Moon with the lore of the constellations not far behind. Virgo rising in the spring heralded the growing of crops and her setting in the fall gave notice that the year's growth was over.

The plaque affixed to the façade of the Messiers' house in Badonviller. It reads: *Left.* Astronomer of the Navy, Member of the Bureau of Longitudes. *Right.* Member of the Academy of Sciences, then of the Institute and, all of the academies of Europe. *Bottom.* Born in this house on June 26, 1730. Passed away at the Hotel Cluny Observatory in Paris on April 12, 1817.

No one knows for certain what sparked Messier's early interest in astronomy, but there were two well reported events during his youth that piqued the curiosity of almost everyone. In March of 1744, when he was just 13, he saw the spectacular six-tail Chéseaux[1] comet. Reportedly, this comet appeared brighter than Jupiter and its reflection could be seen in bodies of water. Later, at the age of 18, he witnessed, from his hometown, an annular solar eclipse on June 25, 1748. Charles must have talked excitedly to his brothers about these incredible astronomical events, his face alight with that childhood glow of awe and wonder.

When Charles awoke each morning and looked out the front windows of his house, he would have seen, directly across the way, an old Protestant church with its accompanying tower, a remnant from a time when a Calvinistic movement had swept the area. However, the structure that stood out the most, just a little farther away and higher at the top of the town's hill, was St. Martin's church whose domed bell tower stood 130 feet high. Charles wrote in a journal that "its three bells of a rather beautiful size could be heard one and a half leagues[2] away when all three were rung together, and this at a time as calm as the night."

---

[1] Philippe Loys de Chéseaux (1718–1751) was a Swiss mathematician and astronomer. He co-discovered the great six-tail comet Klinkenberg-De Chéseaux in 1743. More importantly to this history, he also catalogued 21 deep sky objects by 1746. This included 14 objects that were later to become objects in Messier's catalogue. In 1746, Chéseaux gave his list to Reaumer, who presented it to the French Royal Academy of Sciences on August 6, 1746, but the list was never published. It is very likely that Messier had access to this list when searching the heavens.
[2] A length between 2.4 to 4.6 miles, depending on country.

Charles was fortunate to have a brother like Hyacinthe, who taught him a variety of practical administrative skills, including the benefits of keeping good records, following through on assignments and paying attention to detail. Although Hyacinthe was training his younger brother as a future understudy with a procurator, he was also imparting skills that would be invaluable to an observational scientist. Charles was particularly noted for his good handwriting, an ability to draw and steadfastness in keeping a journal.

Hyacinthe watched his brother grow and mature into a man. Then, coincidently, events came together in 1751 that would send Charles to Paris and Hyacinthe to Senones. After several years of mounting political discord amongst kings and the local lords, an agreement had been struck: The eastern part of the greater Salm area would be joined to Germany while the remaining French territory would be reorganized. Hyacinthe remained loyal to the Princes of Salm, and moved to their new administrative center, his mother's hometown of Senones, where he eventually became the general tax collector. However, during this reshuffling, he became concerned about Charles, now 21, who needed a job, which had become scarce with the changes. Therefore, he turned to a family friend, Abbot Thélosen, for a favor. Thélosen had helped with the 1751 agreement negotiation and was willing to find something for Charles during his Paris visits. Two suitable positions were open. One was at the office of a procurator and the other was with an astronomer. Thélosen wrote Hyacinthe, who discussed it with his administrator-friend Bilistein. They came to the conclusion that the position with the astronomer would be more advantageous. Apparently, Charles had little say in this, if any at all. He writes in his diary, "...they decided together that I should accept the...position [with the astronomer].... Having decided on this, all I thought of was my departure. Eight days went by, and the ninth, which was a Thursday, 23rd September 1751, at nine o'clock in the morning, I left Badonviller."

Personality

What was Messier like as a person? As a youth and young adult, I think the word "nerd" would best describe him. I don't say this disparagingly, but use it as a descriptive "snapshot." I was a nerd (my wife says I still am) and so were my friends. Nerds often grow up to be doctors, lawyers and scientists. They tend to be inquisitive and self-absorbed, having a reflectiveness that often develops into a behavior leaning toward introversion.

I have been around amateurs for years and I know that many of you reading this will have personalities similar to Messier's.

Historical innuendoes point to Messier as somewhat quiet and maybe reserved, if not a little shy. Although this may have been his tendency, it certainly

Top Right. Hyacinthe's house in Senones is adjacent to the town's "old" administrative center. Messier visited him here for a few months in 1772, after the deaths of his wife and son. During this respite, he determined the town's latitude and longitude. *Bottom*. The living room in Hyacinthe's house (located to the right of the front door) remains unchanged. This house is privately owned by Roger and Paulette Leboube who have taken an interest in its history and so graciously gave me a tour. Much of this beautiful house is original and in remarkable condition, including the basement.

Paris, France and Life during the 1700s

Paris invokes images of the Eiffel Tower, Notre Dame Cathedral and the Champs Elysées, the grand avenue that leads to the Arc de Triomphe. With these in mind, it is easy to forget that Paris has over 2,000 years of history.

The site of Paris was originally settled by the Romans, circa 300 B.C., at the Parisian Basin, a once swampy area around the Isle de la Cité, that is, the larger Seine river island where the Notre Dame Cathedral was built. Natural springs facilitated development as well as prompting the construction of those ubiquitous Roman aqueducts and baths.

Paris steadily grew and by the end of the 1700s, had become a large European capital with a population of 550,000. Today, this figure is closer to 12,000,000.

The 1700s were ushered in during the reign of Louis XIV. Although he was known for his flamboyant and lavish lifestyle, France did prosper under his rule. And even though economic conditions improved steadily throughout his reign as well as his successor Louis XV's, the first half of the 1700s were spotted with long and sometimes violent strikes. Most of the population saw their situation deteriorating, while some of "la bourgeoisie," that is, working-class entrepreneurs, became rich. And class stratification had gone too far.

The Notre Dame Cathedral, completed in 1345, sits on the Isle de la Cité (City Island), the larger of the islands in the Seine.

The poorest peasants, who worked as sharecroppers, were heavily taxed and had no rights, while the rich aristocracy was by far the most privileged and taxed the least.

Consumer pricing under the last king of France, Louis XVI who took the throne in 1774, decreased, but so did production. A general economic slump in 1778, combined with a bad harvest subsequently created an agricultural crisis that sent the price of staples like bread through the roof. The economy improved little over the next 11 years, exacerbating conditions for an inevitable revolt.

The French Revolution began with the storming of the Bastille fortress in Paris on July 14, 1789. It was a consequence of extreme class stratification, royal over-expenditure and economic woes. Even the American Revolution spurred it on. The rule by monarchy and nobles was replaced by a National Assembly that was more representative of the people. The new government, however, became divided, opening the doors in 1799 for the successful military general Napoleon to seize control. Ironically, the French accepted Napoleon as a new monarch because he ostensibly kept the democratic ideals and reforms from the revolution. His rule ended in 1814 as a consequence of his grandiosity and overextended conquests. By this time, most of Europe had united against him, while the people of France were ready to move on.

Few of us living today could identify with living conditions during the 1700s. No one enjoyed common "luxuries" that we take for granted. For example, by the end of that century, only 150 bathtubs were owned in all of Paris. And public services were scant. The streets served as all-purpose garbage dumps. Trash, and worst of all, sewage, were simply thrown out windows onto the streets, leaving the air foul with the smell of human and animal waste as well as rotting food.

Consequently, illness and diseases spread easily, with life expectancy averaging 30 years. Throughout Europe, parents giving birth to a dozen children were lucky to see two or three survive into adulthood.

The majority of Parisians lived in poverty. The poorest areas were lined with tall, ramshackle housing units often built so close together that sunlight rarely reached their dangerously narrow streets, where pedestrians were ever vigilant to avoid being run over by passing carriages or horses.

The city was divided into districts. The Châtelet district (translated as "castle," referring to the Louvre, which was originally built as a fortress in 1200) just north of the Notre Dame isle reeked with the waste of butcher shops. Even farther north was the Faubourg de St. Denis area (translated as "surrounding vicinity of St. Denis Church"), which was a poor neighborhood where chil-

Most churches survived Paris' face-lift in the mid-1800s, including St. Etienne du Mont, where Charles and Marie were married.

dren ran around naked, with the luckiest dressed in rags. A few miles east, past the then-intact Bastille fortress, the shops of cabinet makers concentrated around the streets of Faubourg de St. Antoine. The area south of the Seine and the Notre Dame isle was more upscale. Messier's residence and observatory at Hotel Cluny lie in its "Latin Quarter," replete with students and colleges.

Half a mile west, centered about the Abbey of St. Germain des Prés, was the Faubourg de St. Germain aristocratic district. And just a little farther south, past the Luxembourg Palace, stood the Paris Observatory.

Much of Paris' present-day aesthetic beauty and infrastructure is due to the efforts of the architect Baron Haussmann, who gave the city a major face-lift in the 1850's. Large areas of the city were completely demolished for residences of the rich, displacing the poorer occupants to the banlieu or city outskirts. But the demolition enabled the construction of badly needed sewer and water systems. It also made room for numerous parks and gardens and the city's signature, tree-lined boulevards flanked by graciously wide sidewalks.

Messier journeyed to Paris in 1751 when he was just 21. And, like those who visit the capital city today, he was mesmerized by the endless myriad activities and the unfathomable grandeur that make this city truly magnificent.

Pont-Neuf, the bridge farthest west on the Notre Dame isle, has a park next to it and is where Messier watched fireworks.

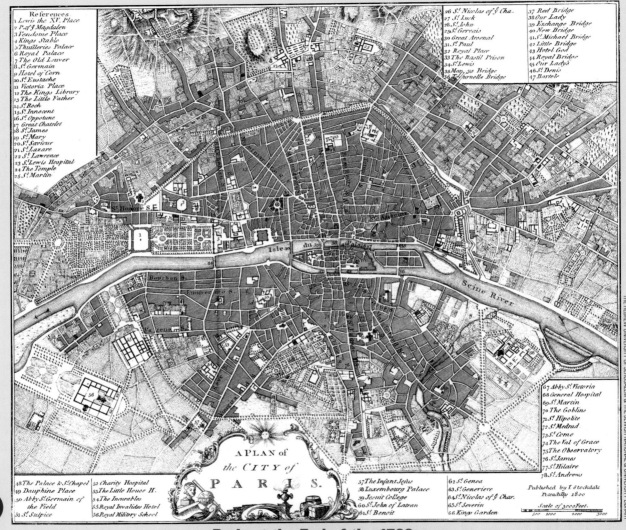

A PLAN of the CITY of PARIS.

Published by I. Stockdale Piccadilly 1800

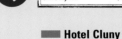

Paris at the End of the 1700s

- ▬ Hotel Cluny
- ▬ Paris Observatory
- ▬ College of France
- ▬ Luxembourg Palace
- ▬ Navarre College
- ▬ Pont-Neuf
- ▬ Notre Dame Cathedral
- ▬ Louvre — Academy Meetings

THE HEBREW UNIVERSITY OF JERUSALEM & THE JEWISH NATIONAL & UNIVERSITY LIBRARY

did not dictate his actions. For example, overall he was very proactive in reporting his observations as well as seeking recognition for his discoveries. It was he who applied for and was accepted as a member of various scientific academies throughout Europe.

For the most part, it appears that Messier was not overly outgoing, even to those whom he knew, and he may have seemed aloof to those he did not know. I would also go so far as to speculate that unless he needed someone to help or support him in his work, he would not go out of his way to foster a relationship or stay in contact with them. However, despite this behavior, there is little doubt that he would have been pleasant to be around. I would bet that if you had been Messier's work-friend, he would have been genial, warm and even light-hearted. However, he would not have been any type of practical joker. The driving force in his life was his work, which made him serious and on task most of the time.

A sense of Messier's inner self can be gleaned from an April 1798 observing note in which he mentions his brother Nicholas-François' visit on the very night that he discovered a comet. Accompanying Nicholas-François was his 14-year-old daughter Joséphine. Messier had not seen this brother in 26 years, and did not know that he had a niece. He indicated almost embarrassment for being an absentee brother; however, he was happy to share the joy of having just discovered a comet with family members. Although this note was brief, it revealed both an inner warmth and a conscience. And it was a rare insight, for his personal thoughts rarely surfaced in any of his notes or documentation.

This note also makes it clear that Nicholas-François was not a part of Messier's life. On the other hand, for decades Messier kept close, letter-writing contact with his brother Hyacinthe in Senones. In all the information available about Messier, nothing is mentioned about his mother, not even the briefest of a reference. She died in 1765 at Badonviller and it is not known if Messier attended the funeral.

Who today, in the astronomical community, might be like Messier? I believe that David Levy, discoverer of 21 comets, including Comet Shoemaker-Levy 9 that crashed into Jupiter, fits the bill. Like Messier, David has a tireless passion for searching for comets. Additionally, similarly to Messier, he is a prolific writer, sharing astronomy with others. I think their biggest difference would be one of affect: David would come across as more jovial and David likes practical jokes.

A family missed

Messier's shy and introverted side becomes evident in the sheer length of his "courtship" with Marie Francoise Dorlodot de Vermanchampt, which lasted 15 years before their marriage. Marie must have had the patience of a saint. She was the daughter of a nobleman glassmaker from Argonne, a region of France northeast of Reims. Apparently, she enjoyed astronomical observing, but I think she also used it as a means to cultivate a relationship with her hopefully husband-to-be. They were married on November 26, 1770 when he was 40 and she 37, at the still-standing St. Etienne du Mont Church, a five-minute walk from Hotel Cluny (see page 19).

Why did Messier court for so long? Were there any social status issues, monetary support concerns or just plain personal reasons? We will never know for certain, but some evidence indicates that

Common Misconceptions about Messier

Most Common Misconception

Messier catalogued objects that could be mistaken for comets.

More Accurate Description

Messier was an observational astronomer who started a catalogue of nebulae that looked like comets, but soon changed it to a catalogue of general deep sky objects.

Discussion: The notion of this misconception is founded in truth. However, it does not accurately reflect the ongoing intent of the catalogue. Although the original impetus behind it was to list objects that could be confused with comets, it did not take long for Messier to realize that there were very few qualifying nebulae, so he continued to catalogue deep sky objects because he was a scientist and his forte was exploring and recording everything in the heavens.

Messier was one of the first observational astronomers who used a telescope to systematically search the sky, accurately measuring and recording comets, nebulae, star clusters and other astronomical phenomena. When he began observing in the mid-1750s, there were no published catalogues listing deep sky objects, just lists that were not widely circulated. Messier became the first astronomer to combine and verify the objects in these lists, while searching for more.

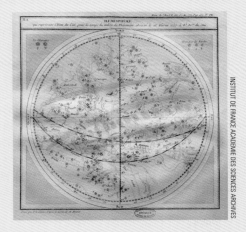

INSTITUT DE FRANCE ACADEMIE DES SCIENCES ARCHIVES

Messier reported on all types of astronomical phenomena. The above chart from 1777 indicates a display of Aurora Borealis. The diagram below from 1790 is devoted to Saturn. Both were published in the annual *Mémoires de l'Académie des Sciences*, an official publication of the French Academy of Sciences.

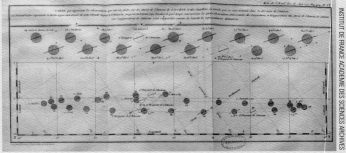

INSTITUT DE FRANCE ACADEMIE DES SCIENCES ARCHIVES

Setting the Record Straight

Second Most Common Misconception

Messier was just a comet hunter.

More Accurate Description

Messier was one of the leading observational astronomers of his time.

Discussion: Messier is appropriately summarized as a great discoverer of comets, but the truth of the matter is that even though he enjoyed searching for comets, he also observed, recorded and reported on every type of astronomical event as well as weather phenomena (see more about weather under the 1776 entry on page 35).

As a member of 17 scientific academies throughout Europe, Messier was recognized as a leader in his field. From 1764 to 1808, he published, in various scientific circulars, descriptions of over 100 astronomical observations and events.

For a period of several years, the frequency of his publications reached one a month. And of his published articles, just a little more than 30% were related to comets. The remaining 70% spanned the arena of celestial events, including eclipses of the Sun and Moon; Jupiter's belts and Galilean moons; Saturn's cloud belts and its rings; occultations of stars and planets by the Moon; the Aurorae Borealis; Moon halos; transits of Venus and Mercury; extreme meteorological conditions; heating caused by the Sun; and of course, observations of deep sky objects.

finances could have played a considerable part in the procrastination. It was probably next to impossible to support a family on a clerk's wages, but easier with the annual salary granted him when he was appointed Astronomer of the Navy, one year after they got married. Messier's marriage may very well have been scheduled on the basis of this pending appointment.

After 16 months of marriage, Marie gave birth to their son Antoine-Charles, on March 15, 1772. This normally joyous event turned tragic, for Marie died eight days afterwards, followed by their son, three days later.

Some psychological textbooks list life's greatest stressors. Ranked at the very top as the most stressful event that can occur, is the death of a spouse or offspring. Messier encountered a double familial death — one of the worst ordeals that can befall anyone. For all intents and purposes, his earthly world temporarily came to an end. A week after his wife died, the very night of his son's death, Messier had nowhere to go, no home with loved ones to return to, so he went out and observed a comet and continued to do so for more than a week. Some might view this as callous, but I think otherwise. His closest loving friendship, that of his wife, was gone and the hopes and dreams that a new child brings, were taken away. His only remaining solace was with friends that would never abandon him, those whom he had spent countless nights with before — the "stars." They provided him with heavenly support, allowing him diversion and even solace to reflect upon his misfortune.

It is extremely unfortunate that both Marie and their son died shortly after childbirth, because I think the experience of fatherhood would have

23

"loosened" Messier up by adding a dimension to his life that only fatherhood brings.

The anticipation of becoming a parent is a mixture of excitement and trepidation. However, for most, the birth itself is an incredible experience, a joyous occasion when time almost stands still. Thoughts are focused on a future family, growing together and enjoying one another for many years to come. All of this was taken from Messier. All of the emotional energy that he had invested, all his thoughts about the future and the companionship that they would provide, died with their deaths. His internal frustration would have grown extremely high from having been denied a major life experience. So, in the fall of that year, Messier did a very sensible thing: he took a three month sabbatical to Lorraine, staying with his brother Hyacinthe in Senones. This gave him time to reflect and heal from his personal tragedy; however, the impact of these events probably weighed upon him and exacted a toll for many years, if not for the rest of his life. He never remarried and I doubt he ever again courted.

Painting a sense of humor

In March of the year 1771, when Messier was finally appointed Astronomer of the Navy, the artist Ansiaume (who studied with the famous pastel portraitist, Maurice-Quentin de la Tour) painted his official portrait. This portrait currently hangs at the Paris Observatory and is shown on

Below. A 270° view from the small square immediately in front of Hotel Cluny. The tower where Messier's observatory was placed is visible. The Sorbonne College is the building at the extreme right and the "gap" to its immediate left leads directly to the College de France where Delisle and Messier had once lived and worked.

page 9. On the back of this portrait, Messier wrote, "This portrait is a good likeness, except that I appear younger than I am, and I have been given a better expression than I have." Did Messier have a sense of humor or was he stating the truth? This statement is indicative of joviality that is probably as much a result of his recent appointment as a reflection of his personality. Additionally, he had gotten married just months before. At this time in his life, he was in good spirits and probably felt on top of the world. However, the statement also indicates his scientific observing ability. The person in the picture *does* look younger than a 40-year-old and there *is* almost a smirk-like smile on his face. Maybe the smirk was the artist's recognition of a whimsical side to Messier that the astronomer did not see in himself.

All in a day's work

Messier worked every day and most clear nights. There can be little doubt that this was of his choosing. One task that he performed three times a day, every day, was the recording of meteorological data.

His position as Delisle's assistant, and later as Astronomer of the Navy, undoubtedly consumed daylight hours. He then spent evenings and pre-dawn mornings studying and searching the sky for comets and other astronomical phenomena. I would not hesitate to say that his work was his life with little time spent on "trivialities" like leaves and vacations.

Even when Messier was away from the "office" and observatory, that is, taking time to roam and explore the streets and surrounding areas of Paris, he could not resist recording in journals what he

Messier's Telescopes

Messier used a number of telescopes throughout his career, possibly a dozen or so. At his time in history, telescopes were described primarily by their focal length, in feet, since the metric system had not been adopted, and with "single" magnification because telescopes were built with an integral eyepiece.

OBSERVATOIRE DE PARIS

Most of his refractors had diameters in the 3 to 4 inch range. The diameters of his reflectors ranged from 6 to 8 inches, but their mirrors were made of speculum metal, meaning metals that can be made reflective, which greatly reduced their performance.

Messier found Halley's Comet using a Newtonian reflector with a 4½-foot focal length and magnification of 60x. This reflector was owned by Delisle and may have been the first telescope placed atop the Cluny tower.

Owen Gingerich, a historical scholar at the Harvard-Smithsonian Center for Astrophysics, reports that one of Messier's telescopes was a 32-foot focal length, 7½-inch diameter Gregorian reflector with a magnification of 104x. He adds that later in life, Messier preferred 3 to 4 inch diameter achromatic refractors with focal lengths of 3 to 4 feet and magnifications of around 100x.

Although the optics of the Gregorian reflector are folded, such instruments were generally bigger, heavier and thus less portable than any 3 to 4 foot refractor. If the Gregorian reflector was anything like the vintage telescope shown on the next page, it may have been the telescope of choice for measuring the celestial coordinates of objects.

Messier discovered many of his comets, including his first and last, with refractors. At one point, he praised a Dollond

It was a Dollond refractor like this one that Messier highly praised, made by the famous London instrument maker John Dollond (1706–1761). Messier seemed to prefer refractors when searching for comets. Today, in England, refractors in general are often referred to as "Dollonds." The building behind the telescope is the Paris Observatory.

achromatic refractor (focal length of 3½ feet and magnification of 120x) by writing that it was "one of the best ever produced by this skilled artist." He probably used these smaller and more portable refractors to observe offsite, in the gardens of Hotel Cluny or, as he reports, in the Louis-le-Grand college, next to the Royal College of France.

The Hotel Cluny observatory was very small (see page 34). It most likely held a telescope that had setting circles built into its mount, a sidereal clock for measuring Right Ascension and a writing surface. Messier used an oil lamp for lighting and to illuminate the eyepiece's reticle. Extra floor space may have been rented under the roof of Hotel Cluny to use as a work area as well as for storage of other equipment and telescopes.

Messier was an excellent observer, but was hampered by limitations in telescope technology and his city observing locations. Also, he was one of the last to use smaller diameter telescopes, since William Herschel and others were starting to make and observe with much larger instruments by the mid-1780s.

Summary of Telescopes Used by Messier

Number of telescopes used: 12 or more.

Objective diameters: 3 to 8 inches.

Focal lengths: Refractors ranged from 1-foot to 30-foot while reflectors ranged from 4½-foot to 32-foot.

Magnifications: 44x to 138x with several around 120x.

Optical designs: Simple refractors (single element for the objectives), achromatic refractors, Newtonian reflectors and Gregorian reflectors.

OBSERVATOIRE DE PARIS

A circa 1700s Gregorian reflector. Like modern Schmidt-Cassegrain Telescopes, the eyepiece is located behind the mirror, at the "back" of the tube. This kind of telescope, with built-in Right Ascension and Declination scales, would have allowed Messier to accurately measure coordinates of celestial objects. Telescopes like this often had a reticle eyepiece that could be illuminated.

saw. This includes writing about the execution of criminals, fireworks seen from Pont-Neuf (see pages 19 and 20) and visiting Versailles. Unfortunately, only snippets from his journals remain, even though he regularly made entries for at least seven years after his arrival in Paris. So, even when Messier was "off," he could not rest his mind. He was the consummate observer because almost everything fascinated him.

Private societies

The Scientific Revolution ushering in our modern technological age started in Europe about 200 years before Messier's birth. At that time, the ecclesiastical hold, which had halted progress for 1,000 years, was loosening, letting scientific investigation bring forth understanding of heaven and Earth, and fueling advancement and a better quality of life. However, this revolution was limited to a small number of people, mostly the "better-off" or society "upper crust" who could afford a good education and had the time, money, equipment and freedom needed for experimentation and exploration.

To facilitate the exchange of scientific ideas and information, academies, societies and institutes were established at the end of the 1600s and beginning of the 1700s throughout Europe, generally one per country. For the most part, these were closed clubs with limited and elected memberships based on some mixture of social status, accomplishments and political connectedness.

These became the clearinghouses for the whole gamut of scientific knowledge, including medical, engineering and instrumentation (like microscope and clock technology). Their meetings were devoted to reporting new findings, presentations and even experiments. Information was disseminated by publishing minutes as well as other literature.

Ironically, at a time when the fastest form of transportation was horse and sailing ship, there was a *considerable* amount of information exchange between these organizations. In fact, the sharing of scientific information was considered so fundamentally important, that it was sometimes passed between countries at war.

Messier was eventually elected into 17 science academies throughout Europe. The first was the Academy of Harlem (Netherlands) in May of 1764, followed by the Royal Society of London that December. There is no doubt that his discovery of six comets by this time as well as his continual reporting and exploration of the sky were factors in these elections.

The French Royal Academy of Sciences was established in 1666. In 1699, Louis XIV gave them a statue and a permanent meeting location at the Louvre (the revolution caused the disbandment of the organization in 1793, but it resurfaced in 1795 as the National Institute of Science and Arts).

Messier was not elected into the prestigious French Royal Academy of Sciences until 1770, and it was the seventh academy to make him a member. He was selected as one of two members to replace the deceased adjunct astronomer Abbot Chappe. His nomination and competition are noted below. Cassini, the other elected member, is Cassini IV, the great-great-grandson of the famous Gian Domenico Cassini.

"On Saturday 30th June 1770, the Academy followed the ordinary procedures leading to the election of two subjects to the appointment of adjunct astronomer, vacant due to the death of M. L. Chappe. The candidates proposed by the

committee were Messier, Cassini, and Duvaucel, among whom the first votes were for M. Messier and the second votes for M. Cassini."

– Account of the sessions of the Academy of Sciences, 30th June 1770

Messier was a fixture at the Academy of Sciences meetings as well as a major contributor to their annual publication, *Mémoires de l'Académie des Sciences* (Reports of the Academy of Sciences). His first article appeared in the 1759 issue. All of Messier's charts shown in this book are from these publications.

Never a stone's throw away

An unusual aspect of Messier's life was the fact that he rarely left Paris. His most significant excursion appears to have been a fourteen-week sojourn on the Baltic Sea aboard the French naval ship *L'Aurore* in 1767, to test and regulate marine chronometers. During these trials, he performed all necessary astronomical observations while Alexandre Pingré, known for his ability to compute the orbits of comets, performed the required calculations.

There is no indication of him attending the funeral of his mother at Badonviller in 1765, the same year that Delisle retired as Astronomer of the Navy. Messier's last excursion outside Paris may have been in the fall of 1770, when he visited his brother Hyacinthe in Senones for three months. It was earlier in that year that his wife and son had died.

Tiptoeing gently

None of us, raised in modern-day America, can relate to the years of turmoil Messier experienced when the French uprising began in Paris at the end of the 1780s, continuing into the 1790s. During these years, heads, including the King's, literally rolled from the guillotine. Messier's friend Jean Bochart de Saron, an ex-president of the Royal Academy of Sciences, died at the guillotine in 1794. The political unrest created massive government confusion, and for a time, Messier's income was cut off — like that of many civil servants. To get by, he was able to borrow from his friend Lalande (see page 43). Although Méchain was an aristocrat, he probably would have escaped any political retribution, but it may have been best that his survey work took him away from Paris and out of the country during these unsettling times. Despite the tenseness of these years, Messier continued to explore the sky, and discovered his eighteenth comet in 1793. Publication of his observations stopped from 1791 to 1796, but resumed in 1797 and continued on through 1808.

Leapfrog

During the 1700s, progress marched forward just like it does today. And, like today, it was only a matter of time before someone came out with something "bigger and better."

Messier provided something "bigger and better" when he published his deep sky object catalogues, supplanting various lists compiled by his contemporaries. Just the act of publishing the catalogues advanced astronomy, in effect giving notice to astronomers that there *are* various types of celestial objects worthy of study and classification, just like flora and fauna. Messier's catalogues and work also inspired others.

Although Messier's last catalogue was published in the 1784 *Connaissance des Temps*, it was printed in 1781. So, it was in December of 1781 that William Herschel was given a copy of it and thus became motivated to compile his own catalogue. Herschel had discovered Uranus earlier that year. Subsequently, he and his sister Caroline developed a routine to systematically search for and record deep sky objects with a 18.7-inch diameter, 20-foot focal length Newtonian reflector telescope. By 1785, just a few years after they started, they had catalogued their first set of 1,000 objects, then another 1,000 by 1788 with 500 more by 1802.

Herschel's efforts effectively put an end to Messier's cataloguing. Messier could not compete with a larger telescope located at a darker site. Nor did he have the time and resources to focus exclusively on cataloguing. In the 1801 edition of *Connaissance des Temps*, Messier comments on and compares his catalogue to Hershel's. Although Messier felt bested by Herschel, the historical irony is that his "puny" 100-object catalogue is still recognized and used today, while Herschel's catalogue has been largely forgotten because it was absorbed into the larger NGC catalogue (see page 52).

A meeting with destiny: The Emperor of France

In 1806, Napoleon (1769–1821) bestowed on Messier the Cross of the Legion of Honor for his lifelong work as an astronomer. Subsequently, Messier wrote an article that was published in 1808, ostensibly stating that the bright comet of 1769 was a portent of the birth of Napoleon. This article created a stir among scientists, who criticized Messier for blatantly propagating pseudoscience.

I don't believe that Messier thought comets were portents of anything. He was an astute scientist and knew very well that comets show up on their schedule — anything happening at the same time on Earth is just coincidence. And, considering the political hysteria, I see the criticism from other scientists as easy potshots. Within the span of a few short years, France went from a monarchy to a constitutional government which was taken over by a military general who was egotistically bent on becoming emperor of Europe. To put it bluntly, Napoleon was a despot, and I have little doubt that it was his entourage who came up with the idea of awarding a medal to Messier in exchange for him writing an article geared to stroking their ruler's ego.

Messier did not get a fancy street sign like Méchain (page 42). And his short, 300-foot-long street, although near the Paris Observatory, is flanked entirely on one side by a prison wall.

Joseph Nicolas Delisle (1688–1768)
Messier's Employer & Mentor

In the autumn of 1751, Abbot Thélosen, while in Paris and acting in the capacity of negotiator between the King of France and the Princes of Salm, met with astronomer Joseph Delisle who needed an assistant. Hyacinthe Messier had asked Thélosen for a favor, that is, if he would be so kind as to find a position for his brother Charles, while in the capital city. Thélosen thus "sold" Delisle on the young man from Lorraine.

The abbot had known the Messier family for years and appreciated their loyalty and service. He most likely gave Charles his highest recommendation, no doubt reassuring Delisle that this youth came from a good family with a good work ethic, was trained as an administrator's assistant, and had good handwriting and an ability to draw. Solely based on his discussion with Thélosen, Delisle agreed to hire the Lorrainian, if he was inclined to accept the position.

Joseph Delisle was born in Paris into an aristocratic family. His father, Claude Delisle (1644–1720) held a royal office, but had also studied law, taught history and geography and worked in cartography. Joseph's oldest brother, Guillaume (1675–1726)

I was unable to locate a picture of Delisle, but I was able to find his name listed on an outdoor plaque in a courtyard at the College of France. Notice that his name has a space between the DE and LISLE, a common practice for aristrocrats before the revolution.

was elected a member of the French Royal Academy of Sciences, but is best known for setting new standards in cartography, eventually receiving the title First Royal Geographer. All three had ties to the Royal College of France.

Although Joseph followed in the family footsteps, his passion was teaching, as well as encouraging and facilitating the advancement of astronomy. He is credited with correctly postulating that Sun halos are caused by light diffracting through water droplets in clouds. Additionally, he was instrumental in initiating the first worldwide systematic study of the 1761 Venus transit of the Sun.

Delisle graduated from the Royal College of France and at one point had the famous Gian Domenico Cassini (1625–1712) as a teacher. In 1718, at the age of 30, he was appointed a professor at the college. In 1724, he was elected a member of the French Royal Academy of Sciences. In this same year, he traveled to London, met Sir Isaac Newton (1643–1727), was made a member of the British Royal Society and received some astronomical tables from Edmond Halley (1656–1742). Delisle's social status allowed him into upper echelon circles, providing him with opportunities to bend ears for the advancement of science. His reputation grew and preceded him, extending as far as Russia. In 1725, at the invitation of Peter the Great (1672–1725), he left Paris for this northern country. Staying there 21 years, he helped establish the St. Petersburg Observatory.

Delisle had a lifelong interest in the acquisition of astronomical and geological manuscripts, especially observational data. While he was in Russia, his collection grew considerably and would become a valuable bargaining tool for him later. Some manuscripts he bought during this time included those of astronomer Johannes Hevelius (1611–1687).

In 1747, at the age of 59, Delisle left Russia and returned to France to become chair of astronomy at the Royal College of France, where he taught mathematics and astronomy. He and his wife, now in their sixties, were childless. They had an apartment at the Royal College, located diagonally "across" the street from Hotel Cluny. After the revolution, the Royal College of France was renamed the College of France with newer construction replacing the buildings where they lived and worked. Today, the College of France

The College of France, known as the Royal College of France before the revolution, is a two-minute walk from Hotel Cluny. Delisle became a professor at this school in 1718 and resumed a position upon his return from Russia. Messier lived and worked here with Delisle during the early part of his career. The school has undergone considerable reconstruction since the late 1700s.

lists Delisle as one of its professors on a wall plaque hanging in a side courtyard.

The Delisle's were cordial to Messier, treating him somewhat like a lost son, and letting him lodge with them at the Royal College until about the time he got married, when he moved to Hotel Cluny.

Delisle, like many astronomers of his time, built and equipped his own observatories. His first, when he was 21, was a small observation post from a dome on vacant crown property — the Luxembourg Palace. Upon returning to Paris from Russia, he again, for a short time, used the Luxembourg Palace until he had a dome constructed atop the front tower of Hotel Cluny, a location convenient to his Royal College apartment.

When Messier started working for Delisle, he initially drew maps. His very first assignment was to copy a map of the Great Wall of China. A few maps later, he worked on a layout of Paris under the direction of Abbot de la Grive. It is reported that Messier was not too happy with the working environment, because his work space was a narrow unheated hallway that got rather cold during the winter.

Although Messier was Delisle's assistant, it was Delisle's secretary Libour who provided Messier's orientation to astronomy, which included the use of instrumentation. Libour drilled into the young man the necessity for documentation and exactitude in measurement. This would have been nothing new to Messier, for no doubt his brother had insisted on similar standards. Libour was apparently pleased with Messier, noting that he was "attentive and interested in all he conveyed."

Undoubtedly, Delisle and Libour felt fortunate to have a worker who was self-motivated, worked long hours and had an insatiable appetite to learn. As Messier became more knowledgeable and skilled, they most likely "rewarded" him with additional responsibilities and challenging tasks. Delisle never realized Messier's true potential because during these times, it was mostly the aristocrats who succeeded since they could afford the luxury of a college education.

In a sense, Messier's "probation" was over in 1753 after he assisted with a transit of Mercury on the morning of May 6th. This event was attended by several scientists. Telescopes were set up aplenty within the gardens of Hotel Cluny. Messier carefully recorded the event, inscribing astronomers' observations, adding descriptions of their instrumentation and making sketches of the transit. He even looked through the telescopes whenever one was not being used by an astronomer. Seeing Messier at work, firsthand, impressed Delisle. He then encouraged Messier to further his studies in astronomy, resulting in the youth attending public lessons taught by Abbot Nollet at the Navarre College (no longer in existence, but just a two-minute walk from the St. Etienne du Mont church where Messier was married). By the end of 1753, Philbert reports that Messier felt that "he was starting to be well trained with the type of work that suited him best."

However, winds were changing. In 1754, Delisle used his collection of geological and astronomical manuscripts to negotiate a position with the Navy. For his transfer of this material to the Navy Map Archives, he was given the title Astronomer of the Navy. Although the reason for this move is unclear, there may have been financial considerations associated with changes at the college. To Delisle's credit, he "brought" his staff with him, securing for Libour a position as Secretary of the Navy and for Messier, the title Civil Servant of the Navy. Obviously, this appointment must have entailed new responsibilities, but apparently it left Delisle and his staff with enough freedom, or the Navy had enough interest in it for them to continue with their regular astronomical work.

The foundation for Messier's observation skills was laid during the years 1754 to 1758. During this time period, he immersed himself in his work, studying astronomy and observing the heavens, meticulously recording everything he saw. Philbert reports that it was on Messier's own accord that he began to systematically look for comets. Little did he know that the years 1758 and 1759 would become life-changing, for the discovery and observation of his first comets were going to be "harbingers" of a stormy sea, one that would permanently change his relationship with Delisle.

About 75 years earlier, Edmond Halley (1656–1742) had become intrigued by the great comet of 1682. Using a new mathematical technique

Hotel Cluny and the Observatory

The observatory used by Messier to find and study celestial objects was atop an octagonal-shaped tower that served primarily as the entrance to what was known as Hotel Cluny. It was built around 1480 as a residence for the Abbots of Cluny, an order of Catholic monks originating from a Benedictine abbey in the city of Cluny (about 220 miles south of Paris) and it remains one of the oldest "residential-type" structures in Paris.

access to the turret, or smaller tower, with stairs leading to the tower's roof where Delisle had a wooden stationary "dome" built. Access to the sky was gained by opening one or more series of shutter-type "doors" that were propped open with iron bars. A drawing of the hotel from this era shows the sides of the "dome" to be flat, following the octagonal shape of the tower, and atop the sides, a roundish pointed roof. Most of the shutter-

Left, center and right views of Hotel Cluny inside its curtain wall (see panoramic view on page 24). This building is just one of three remaining large private "houses" in Paris dating to the 1400s. It was built as a residence for the Abbots of Cluny to support their college which was located on the site now occupied by the Sorbonne.

The building fell into ecclesiastical disuse because the order failed to establish a college, thus making it available for Delisle to construct a tower-top observatory upon his return from Russia in 1747. Although the building was referred to as Hotel Cluny, its function in the 1700s was closer to that of an apartment building.

The building has only one tower, in the front, containing a circular staircase leading to entrances on the ground and two upper floors. On the top floor, or roof-level, there is a second, smaller door to the left of the floor's entrance that provides

windows appear to be located on the sides. The dome was dismantled in the early 1800s.

The diameter of the observatory was small, measuring at most 9 feet. Obviously it was cramped. The only possible means of getting larger items to the top, since its turret entrance was *very* narrow, was by pulling them up with rope.

The single advantage of this observatory was its close proximity to Delisle's place of residence and work. The disadvantages were its small size, blockage from the turret and dome, floor vibration, as well as fireplace and light pollution. When

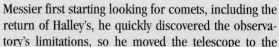

Messier first starting looking for comets, including the return of Halley's, he quickly discovered the observatory's limitations, so he moved the telescope to the hotel's garden, then to the river Seine and finally to the Louis-le-Grand College, next to the College of France. The location of the observatory was a limiting factor that prevented Messier from seeing fainter deep sky objects. Although it is not specifically noted anywhere, one gets the feeling that Messier used the observatory especially for positional measurements but may have frequently used other nearby locations to search for comets.

The book *L'Hotel de Cluny* by Charles Normand, published in 1888, provides a few clues about events at the hotel during Messier's time, including:

1747. Delisle moves his observatory from the dome of the Luxembourg Palace to the Hotel Cluny tower.

1748. Lalande, a student of Delisle, is granted use of the Cluny observatory and boards at the hotel with a counselor (lawyer or officer of the court).

1772. Normand does not indicate when Messier began boarding at the hotel. However, it appears that it may have been in 1772, shortly after Messier's marriage.

1773. Messier observes from towers at the Louis-le-Grand College (immediately adjacent and south to the College of France) offering better views than Cluny.

1776. Extreme cold in Paris from January 30 to February 16. Messier throws water from the Cluny observatory onto the courtyard and notes that it freezes before it hits the pavement. He keeps track of temperatures at Cluny as well as the Paris Observatory. During this cold spell, the hotel becomes Paris' temporary weather bureau, with Parisians stopping by frequently to inquire about the temperature, so much so that Messier lets the doorkeeper disseminate this information.

1789 to 1790. French revolution begins on July 14, 1789. Communique from King Louis XVI's office establishes the ecclesiastical stewardship of the hotel because the church was not actively using the building. Later, in 1798, the Order of Cluny transfers the hotel to the government.

1789. Messier loses his wages and the Navy stops paying the observatory rent. *Continues on next page.*

OWEN GINGERICH

Far left. A closeup of the tower. The turret or smaller tower along its top left side provides access to the roof.

Left. One of the few known depictions of the observatory dome on top of the tower. For an unknown reason, the artist who drew this picture exaggerated the size of the square and the length of the sides of the hotel.

1790. Messier notes that the hotel's chimney restricts his observation of a comet.

1800. After being held for a short time by the government, the hotel is sold to a private party. The Roman baths are not sold and are donated in 1807 to the Hospice of Charenton.

From 1751 to 1798, boarders at the hotel included several counselors, a typographer, a caterer, a book publisher/seller and of course, two astronomers.

Before Delisle transferred his manuscripts and observatory to the navy, Messier called the observatory, "Observatory of Clugny" (note his spelling). Afterward he always referred to it as the "Observatory of the Navy."

Hotel Cluny is now the Musée national du Moyen Âge et Thermes de Cluny (National Museum of the Middle Ages and Thermal Baths of Cluny). Thermal baths were originally constructed on this site during Roman times. The building is located in what is known as the *Left Bank* area of Paris, south of Notre Dame, commonly referred to as the Latin Quarter. Although it is situated on the southeast corner of the Boulevards Saint Germain and Saint Michel where there is a Metro or subway station, the entrance and front of the building actually faces a side street off Saint Michel called Rue du Sommerard, adjacent to the Paul-Painlevé square and across from the north end of the Sorbonne college.

Except for the chapel, the inside of the hotel is not as opulent as its exterior. However, it seethes with robustness and has a Tudor-like charm. This is the second floor room directly behind the tower.

REUNION DES MUSEES NATIONAUX / ART RESOURCE, NY

Today, there is no apparent evidence of Messier's occupation or other astronomical accomplishments at the hotel/museum. Although Cluny has been somewhat modernized and configured for museum displays, only limited interior reconstruction has taken place. The basic shape of the building is an "L" and at the bend, on the second floor, is a beautiful arched-ceiling chapel. The rented apartments were probably located on the first (ground) and second floors, which means that they would have been any of several large rectangular rooms spanning the depth of the building (around 20 feet). The third floor, or area immediately under the roof, may have served as office and storage space for the observatory (it now contains the administrative offices). Part of the hotel was built onto the structure of the original Roman baths which had been at this site since about 200 A.D. A substantial portion of the baths and associated facilities still exists and is part of the museum. The baths are not functional.

From Cluny, it is only a 5 minute walk to the river Seine, 7 minutes to Notre Dame, 2 minutes to the College of France, 6 minutes to the site where the Navarre College stood and 20 minutes to either the Paris Observatory, or the Louvre, which is where the Academy of Sciences meetings were held. Messier did not have far to go to get to anywhere he needed to be.

developed by Sir Isaac Newton for calculating orbits, he discovered that the orbits for the comets of 1456, 1531, 1607 and 1682 appeared to be the same and that if his assumption was correct, it should return in 1758. Halley did not live long enough to see it again, but scientists were waiting. Several, including Delisle, calculated its orbit. Others, including Lalande, even took into account the gravitational perturbations of Jupiter and Saturn. All were hoping to prove their mathematical prowess by calculating the most accurate orbit for the returning comet. Delisle gave Messier the instructions to start looking for the periodic comet in 1757.

It was during this search that Messier came across his first cometlike-looking object, M32, an elliptical galaxy. Then, on August 14, 1758, he found his first comet in the head of Taurus[3]. Unfortunately, it was not the comet that Delisle had him searching for, so Delisle did not make any announcement of its discovery: a practice, to Messier's dismay, that would continued until 1760. On the 28th of this same month, Messier came across a second comet-like-looking object. However, this one turned out to be the Crab Nebula, which eventually would become his first deep sky object catalogue entry.

Finally, on January 21, 1759, Messier found the comet predicted by Halley and subsequently observed it on 13 different days from January 21 to February 13, but there was a problem — it was not

on the path predicted by Delisle. Delisle badly wanted the comet to be found on his path so he could claim credit. Greatly disappointed with his calculations, he forbade Messier to mention his observations to anyone. Furthermore, he dragged his feet by delaying the announcement of Messier's observations/discovery until April. By this time, other astronomers and scientists had become skeptical of the report, because it was already common knowledge that the German farmer and amateur astronomer Johann Palitzch had observed the comet on the night of December 25 and 26, 1758 (word of Palitzch's discovery had not reached Paris when Messier found the comet in January and Palitzch did not immediately recognize the comet for what it was). If Messier had restricted himself to observing the path calculated by Delisle, he never would have come across it. His expanded observing technique bested Delisle's calculations. Although Messier had not been the first to observe the returning comet that became known as Halley's Comet, he would later be recognized as a codiscoverer.

A year passed. On January 8, 1760 Messier made another codiscovery of a comet spotted just one day earlier by Abbé Chevalier. But at last, on January 26 between the constellations Crater and Hydra, Messier made his first original comet discovery. Within one and a half years, from August 14, 1758 to January 26, 1760, Messier had discovered four comets!

There can be little doubt that Messier exceeded all of Delisle's expectations. Although Delisle was probably proud of him, he was most likely a bit jealous too. In a mere nine years, Messier had gone from being a hired assistant to Europe's premier

---

[3] Later, it would be learned that this comet had been observed two months earlier by Monsieur De la Nux, council to the French authority that controlled Bourbon Island, now known as Réunion island, located 500 miles east of Madagascar.

comet discoverer — something that Delisle had never dreamed of. A mere lad from the country — his clerk — was outshining and eclipsing him! Delisle was in a personal quandary about this, and became hesistant to give Messier his "just due," so he dragged his feet, resisting any public acknowledgment of his assistant's accomplishments.

However, the "problem" was that Messier was making important discoveries which could not be ignored. And, to top it off, he had started writing articles for the Academy of Sciences.

Delisle thus had no choice, so in 1760, he finally started to acknowledge the accomplishments not of an assistant, but a fellow scientist, one who had become nothing less than a protégé.

as Astronomer of the Navy, he apparently did not endorse Messier as a successor. This reluctance has all the earmarks of jealousy. However, more could have been going on — at least in Delisle's mind. Although Delisle knew that times were changing and that the days of preferential class treatment were numbered, he probably felt unsure about recommending Messier for a titled position, or found it slightly unpalatable since these traditionally belonged to the aristocracy. He most likely questioned himself repeatedly about this point and even asked himself, "Why does Messier succeed so much?" So, Delisle took the course of least resistance when it came to recognizing and recommending Messier. He continued in that same past vein of procrastination, saying nothing or as little as possible.

The front (pictured) and the back sides of the Luxembourg Palace have domes, one of which Delisle used as a location for observing. Most of the palace was constructed in the early 1600s at the request of Queen Marie de Medici, wife of Henri IV. During the 1700s, this beautiful building and surrounding gardens were vacant crown property. Today, it serves as the seat of the senate. The "back" has a large reflecting pool and a half-mile long grass sward flanked on both sides by Avenue de L'Observatoire, leading to the Paris Observatory.

Messier *was* angry that Delisle had not publically acknowledged his discoveries earlier, but kept his place in respect for his mentor.

In 1765, when Delisle retired from his position

Just three years after retiring as Astronomer of the Navy, on September 12, 1768 Delisle passed away in Paris at the age of 80 from a stroke.

Comets through the 1700s

After the collapse of the Roman empire around 476 A.D., Europe was caught in the clutches of religious fervor thriving on dogma and superstition, dampening freedom, thoughts and ideas.

The continent slumped into 1,000 years of stagnation, which ended at the start of the Scientific Revolution in the mid-1500s that gained momentum from Copernicus' (1473–1543) deathbed book, *De Revolutionibus.* This put forth the concept of a heliocentric Universe, one with the Sun at the center — not the Earth.

Galileo Galilei (1564–1642) followed in Copernicus' footsteps. By 1609, he was making and using telescopes to observe the heavens, finding proof of a heliocentric solar system. Science then took a leap forward at the end of the 1600s when Isaac Newton (1643–1727) invented calculus and provided a theory of gravity.

The 1700s are sometimes called the Age of Enlightenment because they were a period of runaway growth in scientific knowledge, understanding and advancement.

Up until the 1600s, the predominate view on the nature of comets came from the famous 1st century Greek-Egyptian astronomer Claudius Ptolemy (circa A.D. 100–175 from Alexandria, Egypt). Known for his astronomy tome, the *Almagest,* he unfortunately relegated comets to his book *Tetrabiblos,* an astrological treatise. This had the long-term effect of removing these transient denizens from the realm of science and thus serious study. Furthermore, Ptolemy believed that comets were evil portents, which only served to perpetuate them as superstitious objects.

Ptolemy, like Aristotle before him, thought comets were meteorological in nature — sublunar atmospheric phenomena set ablaze

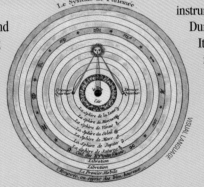

Up through the 1500s, it was commonly believed that comets originated in a sphere of potential fire that surrounded the Earth (band closest to Earth).

within a ring of potential "fire" that surrounded the Earth. However, unlike Ptolemy, Aristotle did not attribute mystical qualities to comets but instead thought that they were purely a weather phenomenon.

Ptolemy's verdict that comets were deleterious was longlasting. Thomas Aquinas (1225?–1274), a Catholic church philosopher, wrote that comets were one of 15 signs of the Lord's coming judgement. Martin Luther (1483–1546) himself equated comets with works of the devil.

The first charting of a cometary path, using a sighting instrument, was made at the beginning of the 1300s. During the 1400s, Paolo Toscanelli (1397–1482) of Italy recorded positional data for six comets, including Halley's. By the mid-1400s astronomers were attempting to measure the distances to comets but the turning point did not occur until the bright comet of 1577, when there was widespread agreement among astronomers that it was much farther away than the Moon. Comets *were* celestial objects!

Although little was learned about the nature of comets during the 1700s, the ability to calculate their orbits confirmed that they obeyed the laws of gravity and were members of the solar system.

Newton originated a method for calculating the orbit of comets, but it was a tedious and circumventive semi-graphical procedure. However, Halley used this method to discover the periodic comet named after him. By 1744 Euler had worked out a purely analytical, but calculation intensive method that was not widely used. During the 1750s, Nicolas Lacaille had improved upon Newton's method, but it was still laborious. Nonetheless, it was used to determine the path of Halley's awaited 1758 return.

It was not until late in the 1700s that Matthias Olber (1758–1840), a physician, finally worked out an "easy-to-use" method for computing cometary orbits which, with some modifications, is still used today.

Pierre François André Méchain (1744–1804)
Messier's Colleague & Catalogue Contributor

Messier met Méchain in 1774, two years after the passing of his wife and son. They quickly developed a working relationship and friendship. I do not think that it would be a stretch to say that Méchain became Messier's best friend. They were formally introduced by Jérôme Lalande (see page 43) who had been a student of Delisle and resided, for a time, at Hotel Cluny.

Méchain was born in Laon into an "economically-strapped" aristocratic family, as his father was a struggling architect. Although he studied mathematics and physics at the College of France, he had to quit because of financial difficulties. Subsequently, he worked as a tutor and for additional money, proofread parts of LaLande's book *L'Astronomie*. In 1774, with the help of Lalande, he became a part-time assistant at the Navy's Depot of Maps and Charts which eventually led to a full-time position as

The Paris Observatory where Méchain was director/manager from 1800 to 1804. The observatory is visible at the very "end" of the main run of the Luxembourg Palace gardens.

"Calculator," where he conducted surveys throughout France and Spain.

When did Méchain's interest in astronomy begin? Did the seed get planted with the proofing of Lalande's book *L'Astronomie*? Since Méchain had some advanced training in physics and mathematics, he undoubtedly had an interest in science, but his finances dictated finding employment. His social status, academic studies and Lalande's recommendation most likely aided him in obtaining a position as a "Calculator" with the Navy. Although his employment took him away from his love of science, Messier rekindled that almost quenched fire. In 1779, five years after they met, Méchain discovered M63, and a year later, his first two comets. Messier and Méchain developed a symbiotic relationship, for Messier needed a friend and colleague as badly as Méchain needed someone to coax him back to science.

One of the biggest complementary differences between these two astronomers was in their math skills. Messier had very little training in this area while Méchain was proficient to the degree that he could calculate cometary orbits. When Méchain discovered his last comet in 1799, Messier provided him with the observational data so he could spend time with orbit calculation.

In 1782, Méchain was made a member of the Royal Academy of Sciences. In 1786, he joined Messier as an associate editor of *Connaissance des Temps* (Messier held this position from 1785 to 1790). Then, in 1787, he, Cassini IV (1748–1845) and Adrien-Marie Legendre (1752–1833), known for contributions to mathematics, collaborated on accurately measuring the longitudinal difference between Paris and Greenwich, England (considered important for cartography), which led the three to visit with William Herschel in Slough, England (a town north of Windsor).

Méchain's meeting with Herschel took place six years after he and Messier had published their final deep sky object catalogue. By this time Herschel and his sister Caroline had catalogued nearly 2,000 deep sky objects, having been spurred on by a copy of Messier's catalogue they received at the end of 1781.

The portrait of Pierre Méchain that hangs at the Paris Observatory.

OBSERVATOIRE DE PARIS

The third and final catalogue listed 103 objects. However, the objects numbered 104 to 110 were never included by Messier or Méchain but were added in the 1900s, mostly from references made in Méchain's lengthy letter of May 6, 1783 to Bernoulli, then in Berlin. This letter "survived" because it was published in the German *Astronomisches Jahrbuch* of 1786 (Astronomical Almanac). In it, Méchain provide data on two comets from 1761, commentary on several nebulous objects that he observed subsequent to the last catalogue publication (namely M102, M104, M105, M106, M107, M108 and M109), orbital information on the new planet Uranus, comments on a transit of Mercury in November 1782, an occultation of the Pleiades by the Moon in February 1783 and finally a brief note on the lunar eclipse of March 1783.

As a Calculator or surveyor, Méchain worked throughout France and Spain. In 1792, a major survey project was initiated to determine more accurate meridians spanning from Dunkerque, France (a northwestern coastal town northeast of Calais) to Barcelona, Spain. Méchain became responsible for field work in the southern portion of this area. This survey started while revolutionary uprisings were still happening. At one point, Méchain and his assistant, Tranchot, were arrested and detained because their instruments were mistaken for weapons of war. Then, while in Spain on the same project, he had a layover because of an injury from an accident. Shortly thereafter, war broke out between France and Spain. Eventually he was able to leave for Italy, but had an extended stay in Genoa, letting things settle down before returning to Paris in 1795. During his absence, and as a result of the uprisings, his family suffered and lost their estate.

Like Messier, once Méchain started hunting for comets, he never stopped and he even took a telescope with him on surveys. While in Barcelona, Spain, he discovered his seventh comet, in January of 1793.

When Méchain finally returned to Paris in 1795, he became a member of the new Academy of Sciences (since the Royal Academy had been disbanded in 1793 because of the revolution) and was elected as one of the original members of the Bureau of Longitudes.

Méchain discovered his eighth and final comet from the Paris Observatory in December of 1799. Founded in 1671, the Paris Observatory had been tasked with improving and/or developing new methods for calculating longitude in order to increase the accuracy of surveys and maps. Since 1671, a suc-

Some street signs in Paris are expanded to include dates, titles and accomplishments of the person named. Méchain has such a sign, posted across from the Paris Observatory. Messier unfortunately did not get this type.

cession of four Cassinis presided as directors. Over this time, the observatory enjoyed royal funding as well as a certain amount of status and independence.

After the dissolution of the monarchy in 1792, a decree in 1795 from the newly formed National Assembly placed the observatory under the auspices of a new organization, the "Bureau of Longitudes," charged with astronomical research to advance science, navigation and geography. Four astronomers sat on its paid, ten member board, including Lalande, Méchain and Cassini IV. Messier was chosen to replace Cassini IV who had resigned after attending just a few meetings. Lalande became the director/manager of the Paris Observa-tory in 1795 followed by Méchain in 1800.

Méchain's last survey became his undoing. Greatly dissatisfied with a survey result from Barcelona, Spain, to the point of becoming depressed, he petitioned, with difficulty, to resurvey the area, all for a 3" difference in latitude, about 300 feet. He left on April 26, 1803 and 17 months later died from yellow fever and exhaustion at Castellón de la Plana, Spain (an eastern coastal town just north of Valencia), on September 20, 1804. Ironically, Méchain had not made any mistakes. The small difference that worried him was a combination of very small effects which, at that time, were beyond his control.

Joseph Jérôme Lefrançais de Lalande
July 11, 1732 – April 4, 1807

Lalande was born in Bourg-en-Bresse, France. He initially studied theology, then law and finally astronomy and mathematics. It was in Paris, while studying law, and lodging at Hotel Cluny, that he became fascinated with astronomy while engaging with Delisle at the hotel observatory.

Although astronomy became Lalande's passion, he promoted other areas of science and was a strong advocate for atheism.

Throughout his life, he craved attention, even resorting to grandstanding. However, he was a "fair" man, able to take criticism, since this also put him in the limelight. In 1773, he craftily stirred the pot to fuel controversy about a comet hitting Earth. At another time, he ate spiders to show that they were harmless creatures. Undoubtedly, Lalande would have fit right into 21st century life.

Lalande was a prolific writer, authoring several books on astronomy and penning numerous articles — many published in the annual *Mémoires de l'Académie des Sciences* (see pages 28 & 29).

OBSERVATOIRE DE PARIS

Lalande reputedly had a wild side or showman-like quality which certainly comes across in this portrait. He stood less than five feet tall, was considered ugly, and pushed an atheist agenda, eventually clashing with Napoleon when the ruler backpedaled on his acknowledged godless beliefs to receive the Pope's approval.

In 1791, he was elected head of the College of France, at which time he opened the school to the enrollment of women. Although he had a propensity for young women, he was attracted to intelligent women, and firmly believed that women deserved recognition for their accomplishments.

Lalande was financially well-off. Not only did he help Messier during the revolution when his salary was suspended, but in 1802, he instituted the annual Lalande Prize for most significant contribution to astronomy.

Other accomplishments and honors of his include:

1753. Elected a member of the French Academy of Sciences.
1757. Calculated the orbit of Halley's Comet with Clairaut, D'Alembert, Euler and Lepaute.
1760–1775 & 1794–1807. Editor of *Connaissance des Temps*.
1762. Appointed astronomy chair at the College of France.
1795. Appointed a member of the Bureau of Longitudes.
1801. Completed a catalogue of 47,000 stars.

The Mystery of the Missing Double Cluster

How could Charles Messier, an astute observer,
forget to include the incredibly fabulous Double Cluster in his catalogue?

If you are relatively new to observational astronomy, your first impression may be that the Messier catalogue is an all-inclusive list of the brightest clusters, nebulae and galaxies visible from the northern hemisphere. This is pretty much true, but there is one notable exception. The beautiful, large and bright Double Cluster in Perseus is missing from the catalogue! With designations of NGC 869 and NGC 884, the Double Cluster was never listed by Messier and the begging question is, why not? How could one of the most exquisite sights in the northern hemisphere, in league with M44 and M45, be overlooked? Either "half" of the Double Cluster could have been a Messier object by itself, but together, they make a whopper. What happened for it to be excluded?

Unfortunately, there is no definitive answer to this question. And, there is just scant historical information shedding any light on this enigma. Perhaps the only person who could settle this is Messier himself, or Méchain. Without any hard facts, we can only speculate on the reasons for its conspicuous absence.

The Double Cluster is a spectacular sight, especially when both "halves" are viewed in the same eyepiece field of view.

N. A. SHARP/NOAO/AURA/NSF

Ironically, the Double Cluster has been known since antiquity as a nebulous mass, because it is visible, under reasonably dark skies, to the naked eye. The Greeks made this nebulous mass the sword hand of Perseus, and this hand had been drawn at this position on charts well into the 1800s.

Messier saw his catalogue as a work in progress and not an end unto itself. The last two catalogues, printed in 1780 and 1781, grew in size, listing new objects. There can be little doubt from Méchain's 1783 letter to Bernoulli, describing other objects, that a "short list" of objects existed for inclusion in future publications. However, it was never to be.

In December 1781, William Herschel was given a copy of Messier's catalogue and was inspired to create his own. He and his sister Caroline worked together, using an 18.7-inch diameter reflector to methodically scan the sky for deep sky objects, effectively putting an end to Messier's cataloguing, for they had catalogued 1,000 objects by 1785 and 2,000 by 1788.

On early celestial charts, the sword hand of Perseus is always drawn at the location of the Double Cluster, or "nebulous mass," based on the description in the *Almagest*. The color chart, from 1720, indicates this position with a star. The larger chart, drawn by Messier in 1790, indicates the nebulous mass by χ (Greek letter Chi) but its symbol is peculiar, compared to that of the other stars. In a February 2003 *Sky & Telescope* article by O'Meara and Green, they indicate that χ marked the Double Cluster on charts as far back as Bayer's atlas from the early 1600s.

Adding the Double Cluster
as M111 & M112

In this book, I have decided to honor Messier by adding the Double Cluster as M111 and M112. My reasons for doing this are explained below.

As you, yourself, continue to explore the sky, you will encounter clusters and galaxies that rival, or trump some of the Messier objects. And even when you come across these objects, you will say to yourself, "How did Messier miss this one?" Many of

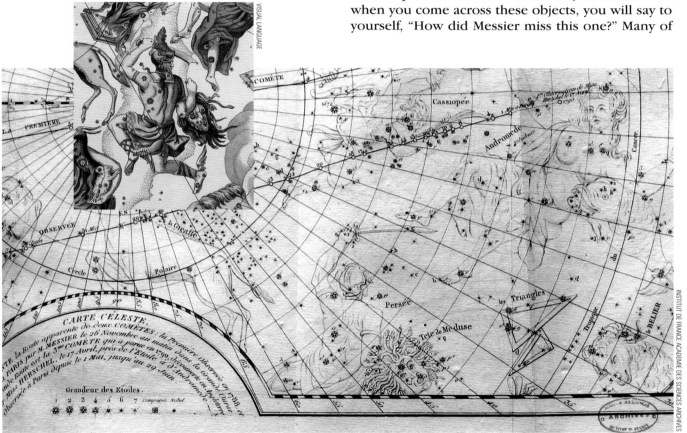

A Ficitional Conversation with Messier about the Double Cluster

Ken Graun: I am honored to meet you, Monsieur Messier. You were my first choice of a historical astronomer whom I wanted to talk to.

Charles Messier: I was an ardent observer and a good one at that but, I lacked the mathematical training of colleagues like my friends Méchain and Lalande.

Ken: That does not make your contributions any less significant. During my time, more professional and amateur astronomers know about your accomplishments than those of any of your colleagues, who were able to afford college.

Messier: (With an astonished expression) I am flattered... and they know me because of my comets?

Ken: You might find this ironic, but it is your catalogue of objects that is well known today and used throughout the astronomical community worldwide. All astronomers, professional and amateur alike, use your catalogue numbers for designation of these objects, but we put an M in front of each number to honor you and to signify your catalogue. Your list of objects is often the first list observed by new amateurs. I almost don't want to say this, but your legacy has more to do with your catalogue than with comets.

Messier: (Blushes a little and looks down) I can't believe this. You do honor me. However, I am a little puzzled as to why I am not known more for my comets. Did Pons[1] surpass me and get all the recognition?

---

[1] Jean Louis Pons (1761–1831), originally hired as a doorkeeper for an observatory in Marseilles, France, is credited with discovering at least 34 comets.

Ken: Yes and no. Pons did discover more comets than you, at least 30, but he is not as well known for his comet discoveries as you. We don't pay too much attention to past comet discoverers because we know what comets are today.

Messier: Ah Dieu! (Oh God!) Please tell me about this! I spent countless hours not only searching for these wonderful objects but pondering their very nature too.

Ken: Comets are simply one type of leftover material from when our solar system formed over 4 billion years ago. They are giant chunks of ice and other chemicals, tens of miles across. As they get close to the Sun, they warm up, which causes the frozen water to change to vapor. The head and tail of a comet are nothing more than water vapor being pushed back by pressure forces from the Sun.

Messier: This is utterly fascinating and overwhelming. I could ask you a hundred questions, but I know you want to ask me something.

Ken: Yes. Why is the Double Cluster, that is, the nebulous mass on the sword hand of Perseus, not in your catalogue? (Hesitatingly) You do know of this cluster?

Messier: Of course I know of the beautiful side-by-side star cluster and observed it many times. So "Double Cluster" is its name — very appropriate. I first learned about Perseus' fuzzy hand from Abbott Nollet when I attended his course at the Navarre College. It's in the *Almagest*. However, I also knew about it from a list of De Chéseaux that Monsieur Delisle, my employer, had in his possession. I worked pretty methodically when I explored the sky. I remember the first time I came across the Double Cluster on a search. I made a journal and mental note to go back for a more detailed look but it got cloudy for several weeks. By the time the nights cleared, I could not calculate its coordinates from the observatory because it was too far west, blocked by the turret and dome.

Throughout my observing days, in order to work efficiently and reduce errors, I did not like to step outside my routine, so I thought I would get back to the Double Cluster on a future pass of the area, but it never happened, for one reason or another. That was a very busy time in my life. I was observing, writing, attending Academy of Sciences meetings, preparing charts, and then there were my duties with the navy.

If I'd known that this would be a future concern, I would have sidestepped my regular searches and catalogued the object, like I did when I discovered a new comet. I did plan to publish additional catalogues. A future edition would have listed the Double Cluster, but I stopped after Herschel compiled a catalogue of 1,000 objects by 1785 and even more thereafter.

Ken: Don't be too hard on yourself. I and many others admire you for all your accomplishments, which are nothing short of stellar — no pun intended. You are very well known in the scientific and especially the astronomical community. Unfortunately, most of your journals have been lost, so this specific journal entry is no longer available.

The tower's turret and the observatory dome obstructed part of the northern sky. In this graphic reconstruction, the red band indicates the arc where the Double Cluster's coordinates could be measured, thus restricting this activity to specific times of the year. The gray area around the turret indicates the *minimum* obstruction from the dome.

Messier: Thank you, Monsieur Graun. You are kind. But even if my notes were lost, you can find evidence that I knew of the Double Cluster from chart drawings of Perseus that were published by the Academy of Sciences. I always drew his sword hand at the "nebulous mass."

Ken: Did Monsieur Méchain know about the Double Cluster?

Messier: We talked about it at one point but he was very busy, if not overwhelmed, with his surveying. It was very time-consuming to add a new object to the catalogue because its coordinates had to be measured. And this could only be accomplished, using a few telescopes, at my observatory or at the Paris Observatory. Because of other priorities, it was not always possible for him or I to determine coordinates. Méchain pretty much did as much as he could given his circumstances. And he did well, discovering many comets and helping with the catalogue. He forgot about the Double Cluster too. (Pause) I missed him in my later years. He was a good friend.

Ken: I am sorry for his early loss. Friendships are what give dimension to our lives. Thank you for sharing all this. (Pause) Real quick. Did you really believe that the comet of 1769 was a portent of Napoleon?

Messier: That is a story unto itself . . .

my observing notes contain the short sentence, "Could have been a Messier!"

Plainly, there are a host of NGC objects that could be "added" to the Messier catalogue because they are equivalent to the famous list of objects. Why not just expand the catalogue and add them? A major problem would be drawing the line in deciding what gets added and what doesn't. For example, I have observed several hundred deep sky objects with my 4-inch refractor — which ones would be chosen? But more importantly, adding objects in any great quantity defeats the purpose of Messier's short but succinct list.

How does the Double Cluster fit in here? Well, there is a huge difference between the Double Cluster and all the rest of the Messier "could-be's." None of the "could-be's" *are* in the same league as the Double Cluster. In fact, few of Messier's catalogued objects *are* in the same league as the Double Cluster. Additionally, none of these other objects have a history that takes them back to antiquity.

Why even add the Double Cluster to Messier's catalogue? Why not just leave the catalogue alone? Adding the Double Cluster completes Messier's catalogue, capping it for the kind of catalogue that it represents, a catalogue of the biggest and brightest deep sky objects that can be seen from mid-latitudes of the northern hemisphere. Without the Double Cluster, the catalogue has a glaring and embarrassing omission, an obvious loose end that leaves it open-ended and confusing, especially to new amateurs.

Throughout the 1900s, objects numbered 104 to 110 were added to the Messier catalogue to give Messier credit for everything he found — to give him his "just due." I am now honoring him for something that he probably knew. I will not dispute the fact that objects 104 to 110 were added because historical evidence indicates that Messier or Méchain had direct knowledge of these objects. I am not advocating that M111 and M112 be "official" entries, as 104 to 110 *eventually* became in the 1900s, but instead honorary entries that allow completion of a catalogue with a glaring omission.

Would Messier have approved of adding objects 104 to 110 to his catalogue as well as the Double Cluster? The answer to this would be a resounding "yes" based on statements he made in the 1801 almanac *Connaissance des Temps* (Knowledge of the Times).

In this edition of the annual almanac, Messier described some of the original impetus behind his catalogue. Although his tone comes across as a bit defensive, most likely from residual jealousy of Herschel's catalogue, he justified his objects as different from Herschel's because they were as a set plainly visible with smaller diameter and shorter focal length telescopes sporting wider fields of view — telescopes that he considered ideal, at the time, for hunting comets. He also mentioned that he had discovered other objects and would publish them in a future catalogue, listing them in order of Right Ascension, but this never happened. So even at the late age of 71, Messier was contemplating the completion of his catalogue. Thus, the inclusion of objects 104 to 110 and the Double Cluster posthumously fits in with the general theme of the catalogue: objects plainly visible with smaller telescopes. The Double Cluster *especially* falls into this category.

For more than 30 years, I have pondered the absence of the Double Cluster from Messier's catalogue. However, it has only been in the last few years, after pulling together historical information on his life and visiting related sites in Paris, that I have had a better understanding of reasons contributing to its omission. Below, I have listed my thoughts and findings about this issue.

Evidence supporting the idea that Messier knew about the Double Cluster

1. The Double Cluster has been known since antiquity as a fuzzy patch in the sky because it can be seen with the naked eye. It is described in Ptolemy's 1st century book *Almagest*, under the constellation Perseus, as "The nebulous mass on the right hand."
2. Those studying astronomy in the 1700s knew of the *Almagest*, since it was an astronomical resource book used throughout Europe for more than 1,000 years.
3. Messier drew charts, published in the *Mémoires de l'Académie des Sciences*, showing Perseus' sword hand at the location of the nebulous mass Double Cluster described in the *Almagest*.
4. On a 1790 chart (see page 45) Messier indicated the Double Cluster with the Greek letter χ, a practice of the time. Although the symbol used to indicate the position of the Double Cluster is for a "star," it is drawn differently than the others. Was this to denote it as something other than a star?
5. In the first sentence of his first published catalogue (see page 332), Messier listed "Cheseaux" as an astronomer who had searched for deep sky objects. This acknowledgment most likely indicates that Messier was in possession of De Chéseaux's list of 21 deep sky objects which included the Double Cluster (actually listed as two entries). This list had originally been passed on to the French Academy of Sciences in 1746.
6. Both Messier and Méchain searched the sky around the Double Cluster, as noted by objects M34, M76 and M103. And there are at least two charts, drawn by Messier, showing comet paths which are in the vicinity of the Double Cluster. It is therefore likely that Messier or Méchain had at least swept by it.

Possible reasons why Messier forgot to add the Double Cluster to his catalogue

A. **Procedural.** Except for the hasty inclusion of objects 102 and 103 to meet a printing deadline, Messier and Méchain never added an object to the catalogue without determining its coordinates. For example, he discovered M62 in 1771, but did not add it to the catalogue until he determined its position in 1779. This was not a quick or simple procedure because it required a telescope equipped with setting circles (scales) and a sidereal clock. Messier could have known of the Double Cluster but never gotten around to measuring its coordinates.
B. **Scheduling.** Messier, as well as Méchain, was extremely busy, a factor that could have contributed to him forgetting about the Double Cluster or about measuring its coordinates. Day-to-day activities that kept him occupied included writing, drawing charts, meetings, comet observing, special celestial and terrestrial events, navy duties, observatory maintenance, basic living, personal and health issues. And, let's not forget any "running" interference from political tensions of the time.

C. **Comet search pattern**. It is not known exactly how Messier searched for comets or deep sky objects. However, since his passion was searching for comets, his observing habits were most likely dictated by an efficient pattern to find them. In modern times, comet hunters tend to observe and scan around the setting or rising Sun. If Messier did this, it would have infrequently taken him to the area of the sky where the Double Cluster is located. Did Messier note fuzzy objects while he searched for comets, and go back to them later for a more careful examination and coordinate measurements? Any delay in returning to an object provided an opportunity for the task to get pushed back on a "to do" or priority list.

D. **Obstacles**. The tower turret and observatory atop Hotel Cluny obstructed a portion of the sky, limiting the "windows of opportunity" Messier had for determining the Double Cluster's coordinates (see drawing on page 47). A period drawing which shows Messier inside his observatory suggests that its structure obscured a *substantial* portion of the sky. It is possible that the Double Cluster was just in an unfavorable part of the sky with respect to the layout of his observatory. And cloudy weather or periods of bad weather could have aggravated the situation by repeatedly "pushing" the Double Cluster out of any observing envelope.

E. **Logistical**. Messier's observatory atop Hotel Cluny was most likely the only facility he had for measuring coordinates accurately. Not only was it equipped with all the necessary instruments, but its telescope was already aligned to the North Celestial Pole (a painstaking, time-consuming task to accomplish), facilitating measurement. Histor-ical evidence indicates that Messier did not always observe from Hotel Cluny, but sometimes also from the Louis-le-Grand College next to the College of France which had better, unobstructed views. If he came across the Double Cluster when observing at this location, he would have had to wait until he got back to the Hotel to determine its coordinates, at least a day's time that could open the door to procrastination and forgetting.

F. **Priorities**. Messier obviously had observing priorities, like plotting the positions of comets, which took time away from cataloguing and other endeavors, distracting him from getting back to and possibly making him forget about tasks.

Messier's oversights & mistakes

Messier was very good at what he did, but like any human being, he made mistakes. Here are a few instances where Messier went astray, supporting the idea that his omission of the Double Cluster could undoubtedly be some kind of mistake.

Messier discovered M110 in 1773 and even drew it next to M31, the Andromeda Galaxy, in 1807. However, he never listed M110 in any of his catalogues. This object was added in the 1900s after it was discovered on this drawing.

There were also several positional errors, mistakes he made in measuring the coordinates of objects in his catalogue, namely M47, M48, M91 and M102. It took some modern-day sleuthing and suppositions to determine the best candidates for these object designations.

Evidence lost?

If Messier had knowledge of the Double Cluster, he probably documented this fact sometime during his life, but this information is probably lost, along with most of his journals. It is remotely possible that evidence may be found in existing documents, but they would have to be scoured closely by someone fluent in French.

Final thought

Historical pointers indicate that Messier should have known about the Double Cluster. So why was it never included in his catalogue? I think there are two main interrelated reasons: Foremost is the fact that the observatory's structure obscured a significant portion of the northern sky, limiting "windows of opportunity" to determine its position. Then there was his busy schedule, which could easily have prevented him from performing this task, allowing it to get pushed back further and further on his "to do" list, until it was forgotten. I end this section with a final fictional conversation with Messier.

Ken Graun: Monsieur Messier, I have one last question — well, actually a request. In honor of all your accomplishments, would it be okay with you if I included the Double Cluster in your catalogue as M111 and M112?

Charles Messier: I am confused. Why would you enter them as M111 and M112 when the catalogue stopped at number 103?

Ken: Sorry. I did not mention that objects M104 to M110 were added to your catalogue in the last 100 years, because these objects were gleaned from notes, charts and correspondence of yours and Monsieur Méchain's.

Messier: I hope their coordinates were accurately determined! Méchain and I only added an object after carefully determining its coordinates. It is most important that scientific publications be as accurate as possible.

Ken: Yes, the positions of these objects were measured very accurately. That is something that we can do easily today.

Messier: The kinds of objects that I included in the catalogue were those visible with small telescopes. The Double Cluster is exactly the type of object that should be in it, but one which I unfortunately forgot. If it hadn't been for Herschel's work, it would have been included in a future publication. I would appreciate your adding the Double Cluster as long as you accurately determined the coordinates of its halves. I still can't get over the fact that my catalogue is my legacy. I would have worked more diligently to tie up some of the loose ends if I'd known that it would be so well-regarded.

Ken: It is your legacy, and a great one at that. I and tens of thousands of professional and amateur astronomers around the world think highly of it, and respect and appreciate you for your accomplishments. Thank you, Monsieur Messier, for your contributions towards making astronomy the wonderful science it is today.

John Louis Emil Dreyer (1852–1926)

Compiler of the *New General Catalogue of Nebulae and Clusters of Stars* (NGC)
and its two supplementary *Index Catalogues* (IC)

By 1908, Dreyer had compiled the largest catalogue of deep sky objects ever — listing over 13,000 objects encompassing the entire celestial sphere. Today, this catalogue is still widely used and celebrated among amateur and professional astronomers. And it may represent the last of its kind, since modern imaging techniques can now record billions of objects.

Dreyer was born, raised and educated in Copenhagen, Denmark. Upon graduating from Copenhagen University in 1874 at the age of 22, with studies in mathematics, physics and history, he moved to Ireland. Like William Herschel before him, he made Great Britain his home. And like Messier, he left home in his early twenties to start his career.

In 1875, Dreyer married Katherine Tuthill of Kilmore. They had three sons and a daughter. Their son Frederic Charles went on to become a full admiral in the Royal Navy, was knighted, and after WWI, represented Britain in the League of Nations.

Upon arrival in Ireland, Dreyer became an assistant to the fourth Earl of Rosse at Birr Castle.

Dreyer in front of the Armagh Observatory.

It was the third Earl of Rosse who had the "Leviathan," a 6-foot diameter, 58-foot long speculum-mirror reflector telescope built on the castle grounds. Birr Castle is located in the small town of Birr, which lies in central Ireland, 70 miles southwest of Dublin. Dreyer used his time at the telescope to publish a listing of 1,000 new objects supplemental to John Herschel's *General Catalogue*. Four years later, in 1878, Dreyer took an assistant position at the Dunsink Observatory (near Dublin) and stayed until 1882, when he was appointed the fourth director of the Armagh Observatory in Armagh, a city 30 miles southwest of Belfast.

The Armagh Observatory was built in 1790 under the direction of Archbishop Richard Robinson, an independently wealthy and influential countryman who founded charitable and educational institutions. Since the Archbishop's authority was seated at Armagh, he had a personal interest in overseeing the city's development. The suggestion for the observatory reportedly came from its first director, Reverend J. A. Hamilton, an avid astronomer who had a private observatory and corresponded with the Astronomer Royal of England. The third director of the Armagh Observatory was Thomas Romney Robinson, who held the position from 1823 until his death in 1882.

J. L. E. Dreyer

When Dreyer became the fourth director in 1882, his publicly funded observatory was almost 100 years old, with somewhat out-of-date equipment. Funding was scarce and became uncertain because of social reforms taking place. In some respects, one has to wonder why Dreyer took this position, but being the director would allow him to dictate the direction of research. Additionally, the solace of a small observatory probably suited his personality and gave him "space" for his contemplated work.

After his appointment, he continued the push for additional funding initiated by his predecessor and was eventually awarded a one-time, 2,000 pound grant, which he used to purchase a 10-inch refractor and hire an assistant to continue his work in studying and cataloguing nebulae and galaxies.

He first completed the *Second Armagh Catalogue of Stars*, a revision of the original work published by the third director in 1859. Next, he had the intention to publish another supplement to John Herschel's *General Catalogue*. Obviously, Dreyer seemed fixated on expanding existing catalogues instead of generating new ones. The NGC and IC catalogues may never have come about if it had not been for the urging of the Royal Astronomical Society to create a new catalogue. Dreyer took their advice and thus compiled the best-known series of deep sky catalogues. *The New General Catalogue of Nebulae and Clusters of Stars* (NGC) was published in 1888, followed by the first supplement, titled the *Index Catalogue* (IC) in 1895 and finally, the *Second Index Catalogue* in 1908 (see IC in the Glossary). Today, these catalogues are known simply as the NGC and IC catalogues. They are currently available in book form as *NGC 2000.0*, edited by Roger Sinnott and published by Sky Publishing Corporation.

For many years, I held the erroneous belief that Dreyer observed and recorded all the objects in his

53

NGC and IC catalogues. I was almost shaken when I discovered that this was not so. Dreyer compiled these catalogues from existing catalogues and observations of fellow astronomers from around the world, including the United States. He undertook the humongous task of collecting, organizing and standardizing the data. Only a few of the entries are

Catalogues Galore

There are other deep sky object catalogues besides Messier's and Dreyer's. Ironically, William Herschel's catalogue is not widely known because it was absorbed into Dreyer's. However, dozens of specialized deep sky object catalogues exist.

In modern atlases, objects without an M or NGC/IC number are usually indicated with the alpha abbreviation of the cataloguer or observatory followed by the catalogue designation. For example, Robert Trumpler (1886–1956) compiled a catalogue of 37 open clusters, while Per Collinder (1890–1974) catalogued around 471 of them. So, Tr and Cr followed by a number are common designations found in atlases. The designation for many dark nebulae (like those surrounding M7 pictured on page 121) begins with the letter B for E. E. Barnard (1857–1922). Designations are numerous and can get confusing, so refer to any introductory or other explanatory information in atlases. More often than not, you will have to search the Internet for detailed explanations.

his discoveries, found when he was at Lord Rosse's observatory verifying and expanding Herschel's *General Catalogue*.

When observatory funds dwindled and he could no longer afford an assistant, he undertook a project of lifelong interest, that of compiling and publishing the works and history of a fellow Dane, Tycho Brahe. Tycho's original manuscripts had lain unpublished in the Royal Library of Copenhagen for 300 years — these were the documents containing the data Kepler used to formulate the laws of planetary motion, firmly establishing the heliocentric or Sun-centered model of our solar system.

Dreyer and Messier had an unusual commonality — they seldom traveled. Messier rarely ventured out of Paris and Dreyer stayed in Ireland. Dreyer did not even visit Copenhagen to study Tycho's manuscripts, but instead had them shipped to him.

In 1890 he published a biography, in English, of the life and works of Tycho. Afterwards he worked on the complete works of Tycho, a 15-volume set in Latin. Finally, in 1906, he published *The History of the Planetary System from Thales to Kepler*.

Dreyer remained director of the Armagh Observatory until 1916, when he relocated to Oxford, England where he died in 1926 at the age of 74.

His distinctions include winning a Gold Medal in 1874, his last year of school at the University of Copenhagen; another Gold Medal from the Royal Astronomical Society in 1916; and serving as president of this society from 1923 to 1925. He held a doctorate from Belfast and an honorary M.A. from Oxford.

The Compilation of This Book

John Mallas, in *The Messier Album* (published in 1978), used a 4-inch, f/15 Unitron refractor for his observations. Stephen O'Meara, in *Deep Sky Companions: The Messier Objects*, used a 4-inch Tele Vue Genesis SDF f/5.4 refractor for his drawings and observations. And I used a 4-inch Tele Vue 101 f/5.4 refractor for my photography, and a 4-inch Tele Vue 102 f/8.6 refractor for visual descriptions. What's up with 4-inch refractors?

I can't speak for the others, but I chose to stick with smaller telescopes several years ago, as I started writing beginning astronomy books, because I did not want to "out-scope" my reading audience. My initial choice was a higher quality 4-inch refractor because I like good imagery and its smaller diameter would give me similar observing experiences to those who are just starting to observe the sky.

On another note, Messier's telescopes were smaller instruments, so it is only fitting — and a little nostalgic — to explore his list of objects with a telescope similar to the ones he used.

My setup for photographing the Messier objects.

Photos & visual descriptions

I don't have an observatory in my backyard and never will, because I live in the immediate outskirts of Tucson, a growing city with a population currently approaching one million. Although the western half of my home sky is fairly dark, the view is somewhat restricted because of surrounding hills and mountains. For the most part, I pack up my gear and head 30 miles down the road to a friend's house in Vail, Arizona, where I have access to a concrete pad that I can back my truck up to for unloading my gear.

Consistency was the theme throughout this project — from the photographs to the visual descriptions. There were no equipment changes throughout the process. I also purchased a sufficient quantity of film beforehand, because film manufacturers have been changing film chemistry and discounting films at an ever increasing rate since digital cameras have become popular.

A list of my equipment as well as a discussion about it are on the following pages.

Photographic equipment

- Tele Vue 101 f/5.4 flat-field refractor, serial number 1125.
- Vixen 60mm guidescope.
- Santa Barbara Instrument Group ST-4 autoguider.
- Astro-Physics 400 German equatorial mount with tracking encoders for JMI NGC-MAX "keypad."
- Astro-Physics adjustable wood tripod.
- Astro-Physics polar alignment scope.
- Jim Kendrick Studio Power Pack 12V, 33AH battery.
- Taurus Technologies camera-back system (this is the original camera-back which was never advertised).
- Tele Vue 11mm & 20mm Plössl eyepieces used respectively for focusing the telescope and the guidescope.
- Fuji Superia 400 speed negative film (this specific formulation has been discontinued and replaced with another).
- Mechanical stopwatch.
- Hat for exposure control.

Visual descriptions equipment

- Tele Vue 102 f/8.6 refractor, serial number 1069.
- Tele Vue altitude-azimuth Gibraltar mount.
- Tele Vue 19mm Panoptic eyepiece yielding 46x magnification and a 1½° field of view (about 3 Moon widths).
- Hand-held cassette tape recorder.

Discussion. I was very satisfied with my equipment choices. Everything operated and performed well, not flawlessly, but very well.

The SBIG ST-4 autoguider was essential to my sanity. Although I can manually guide on a star, the autoguider made "short work" of taking over 100 photographs. I relish this advancement, especially when I think of Clyde Tombaugh, the discoverer of Pluto, painstakingly hand guiding thousands of photographs during his career to find the ninth planet, and to look for others.

The Taurus Technology "camera-back" was another godsend. Roger Blake of Taurus Technology advertises his off-axis guider in *Sky & Telescope* magazine, but it was his "camera-back," something barely mentioned on this website in 2000, that caught my attention. This camera back is a very simple film holder with a shutter that is opened and closed by hand. However, the innovation of this unit is in the focusing. Roger provides a "focusing unit" that is parfocal to the film in the camera-back. A regular eyepiece is inserted into the focusing unit. When the telescope is focused using this eyepiece, the focusing unit (with eyepiece) is removed and replaced with the camera-back. This guarantees perfectly focused images on the film in the camera-back. Not one of the photos was out of focus.

Photo taking procedure

Where and how did I take these pictures? Most of them were taken at either of two dark locations, about a mile away from each other, in Vail, Arizona, 30 miles from my home in Tucson. A third location, which I used once, was in the Chiricahua National

What is Guiding and What Does an Autoguider Do?

To take lengthy time exposures through a telescope, you *must afix the telescope to some kind of equatorial mount* that has its main axis pointed to a celestial pole. A motor attached to this "right ascension axis" turns the telescope at the same rate that the stars move across the sky. However, every motor does not perfectly match the rate of the moving stars (there are several reasons for this), so a small amount of adjustment is always needed to keep the telescope exactly on track.

Traditionally, a guidescope was used to provide adjustment information. A guidescope is not a finderscope but usually a smaller diameter telescope with a long f/ratio that is attached to the main telescope. A reticle eyepiece is used with it to keep a selected star on crosshairs. If the star wanders off, buttons on a hand controller, attached to motors on the telescope's mount, are pressed to minutely move the mount, steering the star back to the crosshairs. The "old-time" method of guiding was to look through the guidescope every minute or so and nudge the scope, using the hand controller, in one direction or another to bring the guide star back to the crosshairs.

Today, autoguiders do this automatically using modern computer wizardry. The reticle eyepiece is replaced with a CCD imaging device that is electronically interfaced with the motors on the telescope mount. A CCD imaging device works similarly to a digital camera. A guide star is found and focused on the CCD chip. If the guide star moves off the pixel that it's set to, the autoguider sends signals to nudge the motors on the mount to move the star back to its original "specified" location or pixel. It continues to do this, effortlessly and more accurately than any human, for hours on end.

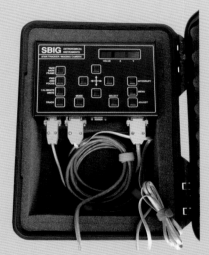

I keep my Santa Barbara Instrument Group ST-4 autoguider in a case for convenient use, storage and transportation. The three cables connect to the CCD head (attached to the guidescope pictured on page 59), the motors on the telescope mount and a 12-volt power source. This type of guider is invaluable for taking pictures with film. Unfortunately, it has been discontinued because most autoguiders are now incorporated into CCD cameras, the preferred imaging "media" used by amateurs and professional astronomers today.

Monument near Willcox, Arizona, 100 miles (as the crow flies) from Tucson.

I arrived at my shooting location around sunset. Usually, one or two of my buddies would already be there setting up for their own observing. I immediately began my setup, which took about 30 minutes. When completed, I waited around to polar align the mount as soon as I could see Polaris and a nearby fainter star needed for the Astro-Physics polar alignment scope. I did not perform a "drift alignment[1]" to achieve the most accurate type of polar alignment but relied solely on carefully using the polar alignment scope. I then waited until it was dark, around astronomical twilight, before taking the first photo.

Each photo was exposed for 20 minutes, plus or minus 10 seconds. I kept track of the time with a stopwatch that hung from my neck. My procedure for these exposures was as follows:

To start an exposure: the telescope, having been set up, had my hat hanging over the front lens, I then...

1. Opened camera shutter.
2. With stopwatch in right hand, lifted hat off telescope objective and held it there for five seconds or so to give the telescope time to settle down from any vibration induced from lifting the hat.
3. Started stopwatch while I moved the hat away.

---

[1] A time-consuming process that achieves the most accurate polar alignment. Basically, it involves observing two stars (one in the east and the other on the meridian) and eliminating their drifting from an eyepiece reticle line by adjusting the azimuth and altitude of the mount.

To end an exposure...

4. Stood to side of telescope objective with hat and stopwatch.
5. Covered objective with hat at 20 minutes.
6. Closed camera shutter.
7. Pressed button to place ST-4 autoguiding unit in "interrupt" or standby mode.
8. Removed camera-back and replaced with eyepiece focusing unit.
9. Advanced film on camera-back.
10. Noted in log next object to be photographed.

To set up for an exposure...

11. Loosened clutch knobs on both axes of mount so telescope would move freely.
12. Used prompts from JMI's digital "keypad" to manually move telescope to next object.
13. Centered object with eyepiece focusing unit, then clamped down clutch knobs on mount. I only had to focus the camera-back once at the beginning of the night, at which time I tightened the focuser down to avoid movement.
14. Using the guidescope with a 20mm Plössl eyepiece, I moved the guidescope, independent of the telescope (see picture caption on the next page), to center a suitable guide star in the eyepiece field of view. The eyepiece was then removed and the CCD head of the ST-4 was placed in the focuser of the guidescope.
15. Pressed several buttons to configure and set the ST-4 autoguider.
16. Switched eyepiece focusing unit with camera-back on telescope.
17. Placed hat over telescope objective lens, then looped back to number 1.

On average, it took ten minutes to set up each 20-minute exposure, so there was a total time of 30 minutes spent on each shot. Most of the preparation time was spent on steps 14 and 15.

I planned the shooting sessions by consulting planetarium software. This allowed me to determine the best time to photograph the target objects, generally when they were high in the sky, preferably near the meridian.

It took 15 outings, totaling less than 100 hours over a period of 29 months, to photograph all the Messier objects. This hour count includes vehicular travel, setup and packing.

All of the photographs were taken between September and the very first week in June because Tucson's summers are normally cloudy and somewhat rainy.

Overall, I had it "easy" taking these photos. The winters of southern Arizona are mild and cloudless compared to other parts of the country. There are few pesky bugs, and dew is almost nonexistent. In southern Arizona, winter lows are in the upper 30s or low 40s, but these temperatures are not reached until early morning. For the most part, winter nighttime tem-

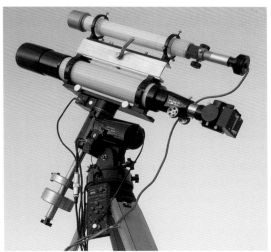

The guidescope is the smaller telescope at the top of the picture that has the CCD autoguider head inserted into its focuser. Inserted into the focuser of the main scope is the Taurus camera-back. Between these scopes is the handle I used to loosen an assembly for moving the guidescope, independent of the telescope, to a guide star. This German equatorial mount has motors attached to both Right Ascension and Declination axes that allow "nudging" in both directions for more accurate tracking.

peratures stay in the 50s. The worst exposure-spoiling weather was occasional wind gusts. However, my mount and scope are fairly robust, so small gusts had no effect on my photographs.

After a night's outing, I had the film processed at a one-hour camera store to get immediate feedback on any problems. In all, I took 134 photos and rejected 25 for various reasons. These two numbers don't subtract to 112 because several Messier objects could often be captured on one frame (for example: M31, M32 and M110). Less than a handful of exposures ended prematurely when the telescope moved far enough in Right Ascension to hit the tripod (I underestimated the distance the telescope moved in 20 minutes). Only one exposure was terminated because of clouds. I also "wasted" one whole night taking pictures when I discovered, upon development, that I had used the wrong film.

One of the two main locations that I used in Vail, Arizona was on state property. It was evident from spent shotgun and rifle shells that people also used the site for target practice. I always went to this location with my "astro" friends. Today, few of us feel safe alone. At this location, it wasn't the remains of ammunition that was worrisome, but the fact that individuals whom I knew nothing about could show up unexpectedly. Since completing this book, I have been observing and photographing only from my friend's house in Vail where I feel safe, whether or not he is home or I am alone.

Memorable moments

On one of the nights, a bolide meteor lit up the sky like a giant camera flash, producing very strong shadows. Three of us estimated it had a magnitude of around −17. Although there was no accompanying sound, it did leave a lingering smoke trail.

On another night, several of us were able to glimpse the shape of the International Space Station as it sped by. We used a Tele Vue 102 on a Gibraltar altitude-azimuth mount (neither were mine) which enabled us to track its rapid movement across the sky.

The most memorable non-astronomical event occurred on September 20, 2001, just nine days after the 9/11 collapse of the Twin Towers. It was around 2:30 a.m. and I was at my friend's house in Vail, packing up to go home. About an eighth of a mile from his house is a north-south 2-lane highway that links up to the interstate. Out of the blue, I heard a long train of vehicles trekking down this road, southbound, away from the interstate. I could see their headlights, but that was all. There might have been 30 to 50 vehicles in the train, all evenly spaced and traveling very close to one another. My initial impression was that they may have been mili-

tary vehicles heading toward the Fort Huachuca army base in Sierra Vista. The odd thing was the sound. A few of the vehicles had motors that cried "souped-up." A second train of vehicles then passed and my curiosity peaked. Some fear began to surface in me about things that go bump in the night. I hurried my packing so that I could get over there and see these things close up. A third train started to go by. Again, all evenly spaced and following each other closely. I quickly got in my truck and drove to the highway. Unfortunately, there were no more "trains" and I was headed in the opposite direction, but there were stragglers. For Pete's sake, they were older vintage hotrod type cars! What the heck were they doing out at 2:30 in the morning, heading south into practically nowhere? I didn't have any idea. Several weeks later, I checked with my friends at a local telescope store. Apparently, there are "hotrod" or vintage car clubs that do these early morning runs. Maybe they are trying to avoid traffic and unwanted attention. I really don't know for sure but I guess they have their reasons. Nearly scared the heck out of me.

Messier Marathons

I have never been one for marathons of any sort. In my opinion, they don't prove anything. However, since I decided to write a book on Messier objects, I thought it would be appropriate to try my hand at a Messier marathon because they have become an annual event in the amateur community.

What is a Messier Marathon?

For those residing in the mid-latitudes of the northern hemisphere, it is the act of viewing every Messier object, except for possibly either M74 or M30, in one night. This can be accomplished only around New Moon during the month of March. Why March? There are two reasons. At this time of the year, the nights are still long, which allows a good portion of the celestial sphere to be viewed. Just as importantly, if not more so, is the fact that the Sun is positioned between the Right Ascensions of 0 and 23 hours, near the circlet of stars in Aquarius, an area devoid, in Declination and nearby Right Ascensions, of Messier objects. So, this is the one time of the year when all the Messier objects can appear in a single

Setup for my first Messier marathon. I found the Messier objects with the refractor on the manual altazimuth mount to the right but had a GO TO telescope, at left, on stand-by, just in case.

night sky. However, M74 or M30 may not be observable because one usually gets lost in the glow of sunlight.

Suggestions for your Marathon

Have fun. First and foremost — take steps to make your Messier Marathon fun! Don't let it be a chore or become a struggle. If you are inexperienced with finding Messier objects, join in with more seasoned observers, but participate with people you know and like.

Go manual. You could easily accomplish a Messier Marathon by using a GO TO telescope, but I encourage you to manually find these objects. This will help fill the hours and you will feel more satisfied with your accomplishment. *Some of the following comments are specific to those wanting to manually find Messier objects.*

Country sky. Seek out a dark, unobstructed sky with as little light pollution as possible. I would not try a marathon in light-polluted skies, because some objects, like the Crab Nebula (M1) and Owl Nebula (M97), simply cannot be seen under these conditions. On the other hand, these same objects

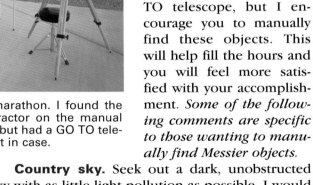

are plainly visible under dark skies. All you have to do is pass by them to catch'em.

Star Charts. Don't get too fancy here. Use a set of easy-to-read, uncluttered star charts, like those in this book. Keep a fainter magnitude atlas on hand just in case you need to verify an object.

Finder. Ideally, equip your telescope with a reflex-sight finder like a Telrad instead of a traditional finderscope. This will speed up finding objects because it provides a one-to-one correspondence between the sky and star charts.

Magnifications. Independent of telescope size, stick with an eyepiece yielding magnifications in the range of 25x to 50x. If possible, use magnifications closer to 30x for locating these objects. Higher magnifications, like 75x or above, yield smaller field of views which makes it more difficult to spot and positively identify objects.

Telescope. Use a smaller, 4 to 6-inch diameter telescope. Bigger telescopes are not better for marathons, especially under dark skies. Smaller telescopes achieve a greater range of lower magnification with larger fields of view. Since the Messier objects are the biggest and brightest deep sky objects, they are hard to miss in a large field of view under dark skies.

My first Messier Marathon

Before sunset on Wednesday, March 13, 2002, I set out on my first Messier marathon. Unfortunately, I was at the tail end of a lingering cold, so I was weak and should have been in bed, but duty called. As I drove to a friend's house in Vail, Arizona, where the skies are unobstructed and fairly dark, I remi-

nisced about the Messier marathon of two other friends the previous year. Although I was with them, I spent two-thirds of the night photographing objects for this book. I, however, got to go home early because I finished a few hours before dawn. Marathons are one of the few dusk-to-dawn events, but with numerous rests in between.

Prior to my marathon, I strategized the evening using a software planetarium program, so I had a prioritized list for observing the objects. I set up two telescopes. The first was a complete GO TO system, on standby, in case I needed to verify any object, and also, as I anticipated, to find the last objects which might get "lost" during early dawn. The second telescope was the one I used, a 4-inch Tele Vue 101 refractor on an altazimuth Tele Vue Gibraltar mount. I was determined to manually find all the objects, using and testing the first draft of the star charts in this book.

I treated the Messier marathon as a "find only" expedition, not an outing for studying these objects. The most I did was look at each object for five to ten seconds, noting their beauty and moving on to the next.

It would probably take two to three hours to manually find and view all the objects if they were under one sky. However, because they are spread across the celestial sphere, it takes the entire night, mostly because you must wait for them to rise in the east so they can be observed.

I did this marathon alone. I set up outside a friend's backyard, on a nice concrete pad that I can back my truck up to. I chatted with him a little when he came out to do some observing in the early evening, and the following morning. Although I stayed out all night, it was comforting to know that safety was just a few feet away.

My observing went smoothly and was uneventful. However, my cold caught up with me in the early morning, so I snoozed in my truck for a few hours between sets of objects. I would have stayed awake if I had not been sick. It was a nice night, with no clouds, a mild temperature and little to no wind. Of course, there wasn't a problem with dew. In fact, there rarely is in southern Arizona. A few hours before sunrise, I heard roosters crow and "houses" starting to stir. That night, the "job" for most of the home dwellers was to sleep, but mine was to stay up and observe a litany of objects from a historically significant catalogue. It seemed ironic that in one night, I saw over 100 objects that took Messier years to catalogue. I can see him smiling now.

The most humorous moment of the marathon was our verbal exchange in the early morning when my friend came out of his house and yelled over his backyard fence.

Friend: "Did you see one hundred and nine?"[1]
Ken: "Yes," I responded. "I saw that one a few hours ago."
Friend: "No," he said. "Did you see one hundred and nine objects?"
Ken: "Oh, yes I did," I replied.
Friend: "Good," he exclaimed.

---

[1] The number of Messier objects normally seen during a Messier Marathon before I added the honorary M111 and M112 to the list.

It was dawn, and obviously my mind was too tired to pick up on the essence of his meaning, but I felt happy when I understood his question, realizing and recognizing the accomplishment of the night. I was surprisingly content about having experienced something that I normally would not do. I highly recommend the marathon as long as you are not sick, and do it with good company who shares the same goal. Do it for your own needs and in honor and remembrance of the hard work of those who came before us.

Observing list

On the next page is a list of the times and order in which I planned to observe the Messier objects. This order should work for the majority of mid-latitude locations in the northern hemisphere. For the most part, I followed my outline, but at times I skipped around, just for the sheer joy of it, and also because I could not resist my anticipation for observing favorite objects.

The most difficult batch of objects to identify is the Virgo Cluster of Galaxies because you have to use a star chart to "verify" each one. Although I observed these objects much earlier than the time indicated on the list, I highly recommend waiting to view them when they are on your southern meridian, because the up and down and left to right movements of the telescope on an altazimuth mount will correspond to the same directions on Charts T & U. If you are using an SCT or refractor with a 90° diagonal, use the mirror-reversed Chart V.

*Have fun, and may you have
clear skies for your marathon!*

Messier Marathon Checkoff List

8 p.m.[1]   74[4], 77, 52, 31/32/110, 33, 103, 111/112, 76, 34

9 p.m.   79, 41, 42/43, 78, 45, 1

9:30 p.m.   36, 37, 38, 35, 50, 46/47, 93, 48, 67, 44

10:15 p.m.   95/96/105, 65/66

10:45 p.m.   81/82, 97/108, 109, 40, 106, 101, 102, 51, 63, 94

11:15 p.m.   3, 53, 64, 104, 68, 83, 5, 13, 92

1 to 1:30 a.m.[2]   Virgo Galaxy Cluster
(or earlier)   85, 100, 99, 98, 88/91, 89/90, 58, 59/60, 87, 84/86, 49, 61

2 to 3 a.m.[3]   12, 10, 107, 80, 4, 14, 57, 56, 9, 19, 62, 29, 39, 27, 71

4 a.m.   11, 26, 16, 17, 18, 24, 25, 23, 20/21, 8, 22, 28, 6, 7

4:30 a.m.   54, 70, 69, 55, 75, 15, 72/73, 2

5:15 to 5:30 a.m.   30[4]

Congratulations. You've completed a marathon of the highest order!

Notes [1] Start when it first gets dark at your location. The actual starting time (as well as the other times indicated) will vary depending on your latitude and location in the time zone.

[2] These galaxies will be on or near your southern meridian at about this time.

[3] Don't dilly-dally from this point on because dawn will sneak up on you fast. Work ahead as much as possible.

[4] Often, either M74 or M30 may not be observed because they get lost in the glow of sunlight. However, they may be visible and located with a "larger" telescope on a GO TO mount.

Discussion on Finding and Observing Messier Deep Sky Objects

Venturing out to find and observe Messier deep sky objects may at first seem a little daunting, and perhaps even cause for some trepidation. However, let me assure you that this endeavor is well within the reach of anyone desiring to do so.

The major difference between finding and observing the Moon or Planets and Messier objects is that Messier objects are significantly fainter. Surprising to many beginners, however, is the fact that the majority of Messier objects appear much larger in the sky than any of the Planets *and* many span an area as large as or larger than the Moon. So, it is not their size but their faintness that initially makes finding and observing them seemingly difficult. Obviously, if all Messier objects were as bright as the naked-eye Pleiades, M45, they would be plain-as-day targets.

Star charts are the road maps to locating Messier objects, while recognizing them is just a matter of becoming acquainted with their characteristics. Observing Messier objects will be good

I enjoy finding Messier objects the most with a manual telescope on a simple altazimuth mount like my homemade 6-inch reflector.

practice for orientating yourself to the general feel of deep sky objects (DSO), preparing you for the natural push toward observing the multitude of DSOs in the NGC and IC catalogues.

Whitewashed skies

The biggest obstacle to observing deep sky objects is whitewashed skies produced by either the light of the Moon or manmade sources. Since deep sky objects are *much* fainter than the Planets, they are easily obscured by a sky flooded with light.

Moonlight. Ever since the telescope was invented, those wanting to observe deep sky objects have had to work around the Moon, a natural "light polluter." Fortunately, its impositions are cyclical, so with amateurs there is always a spurt of activity on the weekends closest to New Moon to take advantage of the long, dark nights without moonlight and glare.

Although the light from a bright Moon whitewashes the sky, I have had success in observing some Messier objects with a quarter Moon, including many galaxies of the Virgo Galaxy Cluster (Charts T, U and V). Under these conditions, try to position your telescope in the shadow of a building to avoid the Moon's direct glare.

Light pollution. Unfortunately, many of us live in cities or towns whose skies are aglow with the spill of unshielded light — a problem that we have to live with. Many Messier objects can be observed in moderately light-polluted skies, but some objects, like M1 and M97, are more

difficult if not impossible to see under these conditions.

If light pollution is a major problem, try observing objects in the portions of the sky that are least affected. However, more often than not, you will simply have to observe at a darker location. I live on the immediate outskirts of Tucson and even though half of my sky is okay for observing, I still have to travel about 30 miles, with all my equipment, to get to reasonably dark skies for more serious deep sky exploration and photography.

The Difference between Observing Messier and NGC Objects

The Messier and NGC catalogues are both lists of deep sky objects. And since Messier's catalogue was the first of its kind, all of its objects are included in the later and more comprehensive NGC catalogue.

For the most part, Messier's deep sky objects are by far the biggest and brightest of the lot. However, there are a number of NGC objects that are in the same league. On average, NGC objects are much fainter and smaller than Messier's objects, thus requiring darker skies and/or larger telescopes with more light gathering capability to be observed easily. All of Messier's objects are easily viewed with a 4-inch telescope, but it would take about a 16-inch telescope to easily see most of the 8,000 NGC objects.

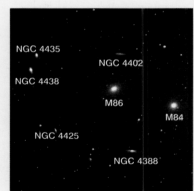

A major difference between Messier and NGC objects is brightness and size, as shown by this mix of galaxies.

Nevertheless, there is one object, namely the planetary nebula, that *does* differ in appearance in the Messier and the NGC catalogues. There are only four Messier planetaries, namely M27, M57, M76 and M97. These planetaries are fairly large and extended. However, many planetaries in the NGC catalogue don't resemble them but instead look like bloated stars or small disks (this is why they were originally given the descriptive name, "planetary"). Some look very starlike in smaller telescopes at lower magnifications. I passed up several NGC planetaries when I first started looking for them because I was expecting something similar to the Messier examples, only smaller and fainter.

Night Vision, Red-light Flashlights & Peripheral/Averted Vision

Night vision is a common term to describe when eyes have become dark-adapted. This refers to the ability of the eyes to see "better" in the dark, to see fainter objects or lower levels of light than normal. All of us have experienced the sensation of walking from a lighted hallway into a dark theater. We can't see the seating until our eyes automatically adjust by opening their irises and letting in more light. It only takes the eyes a minute or so to initially dark-adapt and about 15 minutes or longer for a better, "deeper" dark adaptation. Dark-adapted eyes are essential for viewing deep sky objects.

Bright lights or the use of a bright flashlight to read notes or charts will interfere with your night vision. To preserve your dark-adapted eyes, use a low-intensity **red-light flashlight** for reading charts, writing or working with equipment. I say low-intensity because many red-light flashlights on the market are too bright to use when observing. In fact, some red-light LED lights are brighter than white-light flashlights. The designs of flashlights come and go quickly, so look for one that allows you to vary the intensity of the light. With a variable light source, you will find yourself using the lower intensities once your eyes have become more deeply dark-adapted.

Human eyes behave differently in the dark than they do in daylight because they evolved primarily for daylight use.

For example, very faint galaxies and nebulae will literally disappear if you look *directly* at them through a telescope. This happens because human eyes have a reduced number of light receptors at the center of the retina (called the blind spot) which makes this area less sensitive or less capable of detecting low levels of light. The reason there are less receptors here is because this area also contains extra informational "wiring" that goes to the brain.

So, observers often use "**peripheral vision**" or "**averted vision**" to glimpse faint objects that otherwise cannot be seen when they are looked at directly. In the dark, our peripheral vision is much more sensitive to faint light than "straight" vision.

In addition to this limitation, our night eyes cannot detect the color of faint light, but instead see any faint color as just white light. This is the reason why most nebulae and galaxies appear whitish to the eye, even in larger telescopes.

Using averted vision is a "trick" amateurs have to either glimpse fainter deep sky objects or garner greater detail that cannot be detected when looking directly at an object. Compared to straight vision, the down sides to using averted vision are greater eye strain and degraded focus. Note: See *averted vision* in the Glossary for an additional trick to help detect the very faintest objects — those at the threshold of peripheral vision.

Telescope discussion
salient to viewing Messier objects

The greatest limiting factor to observing Messier's objects will not be the telescope, but the quality of the night sky. All of Messier's objects can be seen with the smallest of modern-day telescopes.

Three basic telescope designs. The three basic optical designs for telescopes are refractor, reflector and catadioptric. Any of these designs will do well for observing Messier objects. The most popular catadioptric design (catadioptric means a compound telescope that is a cross between a refractor and a reflector) takes the form of the SCT, the **S**chmidt-**C**assegrain **T**elescope, originally made popular by Celestron.

Objective size. To view Messier objects, I recommend a telescope with a diameter ranging from 2¼ to 8 inches. Personally, I think it is easiest to find them with smaller telescopes, under 6 inches in diameter, but larger telescopes will provide some advantage in light-polluted skies. Most amateurs go

through several telescopes over a lifetime, so if you start smaller, you can always trade or buy up.

If you get hooked on deep sky objects, the larger the telescope diameter, the better. To pursue NGC objects, you might want to eventually get a 12-inch or larger diameter telescope.

Maintenance, collimation and price. Refractors are practically maintenance-free and easy to use. They rarely require collimation, that is, realignment of their optics. The highest quality refractors, called APOs (which stands for apochromatic, meaning free of any optical defect), provide the best telescope imagery. Unfortunately, good refractors are the most expensive telescopes per aperture inch and are limited to about 6 inches in diameter, which does not make them good candidates for observing the NGC and IC objects.

Newtonian reflectors require occasional to frequent optical collimation — although usually just tweaking. They are most often mounted on manual, altazimuth mounts, which means that you have to

Refractor on a simple altazimuth mount

Schmidt-**C**assegrain **T**elescope on a GO TO altazimuth mount

A Newtonian reflector on a simple altazimuth mount is sometimes called a "Dobsonian"

move these scopes by hand to find and keep objects in the eyepiece field of view. A major advantage of Newtonians is that they give you the biggest aperture for the buck. Sizes range from 4 to 36 or more inches in diameter.

In my opinion, SCTs offer the biggest bang for the buck because their price usually includes a computer-motorized "GO TO" mount that can find and track any object chosen from its hand controller. Per aperture inch, SCTs are mid-priced. These scopes sometimes require optical collimation depending on the amount of "handling" they receive. Sizes range from 5 to 16 inches in diameter.

Telescope mounts

Basically, there are two types of telescope mounts: altazimuths and equatorials. These mounts come in motorized and manual versions.

Altazimuth mounts are the simplest type of mount. They allow the telescope to move up and down in altitude, as well as to be rotated to any compass point or azimuth. Newtonians often have manual altazimuth mounts, while SCTs have a motorized version. See pictures on the previous page.

The equatorial mount (see pictures on pages 59 and 345) offers advantages over the altazimuth, but is a more complex unit. Although there are different kinds of equatorial mounts, all of them have one of two axes pointing to a celestial pole, facilitating the movement of the telescope to "naturally" follow stars across the sky. If you plan on taking photographs requiring extended exposures (a minute or more), you will need an equatorial mount. Although equatorial mounts are usually expensive, they can be found on inexpensive telescopes. Overall, mov-

ing an equatorial mount manually is more awkward than moving an altazimuth mount, especially the closer you point the telescope to a pole.

Bottom line: Unless your mount is motorized, the manual altazimuth mount is the easiest type of mount to use for general exploration of the night sky.

GO TO. Open up any current-day astronomy magazine and you will find an abundance of advertisements for GO TO (also written GOTO) telescopes. What exactly is GO TO? It actually refers to an ability of the mount, rather than the telescope. A GO TO mount has computerized motors that will automatically move the telescope to, and then track, any celestial objects in the sky chosen through a hand controller (you can access various catalogues of objects, including the Planets). Both altazimuth and equatorial mounts can be made GO TO. Some setup is required. For example, GO TO SCTs must be aligned to two bright stars at the start of observing (the hand controller walks you through this process), but once this is accomplished, the telescope easily moves to any chosen celestial object, all night long.

If you have a GO TO telescope, locating Messier DSOs is just a matter of pressing a few buttons. However, if your telescope does not have such wizardry, you must find them the "old fashioned" way, by manually moving your telescope using a finder as your pointer and star charts as your guides.

I love my GO TO telescope mounts and will never give them up, but I also enjoy finding Messier objects manually. I encourage you to find them this way too, becaue it provides practice in using star charts and learning the constellations. Additionally, it gives you a better sense of the lay and movement of the celestial sphere.

Calculating Telescope Magnification & Field of View

Magnification

The magnification of a telescope can be changed by using different focal length eyepieces. **Magnification is calculated by using the following formula:**

> **Focal Length of Telescope ÷ Focal Length of Eyepiece = Telescope Magnification**

DISCUSSION. When calculating magnification, **BOTH focal lengths MUST be expressed in the same unit of measurement** — usually millimeters (mm). If the focal length of the telescope is in inches, change it to millimeters by multiplying it by 25.4 (there are *exactly* 25.4 millimeters to an inch).

The focal length of a telescope is sometimes noted on a label attached to the telescope. Otherwise check the instructions. Length is usually expressed in mm (millimeters).

The focal length of eyepieces is indicated on the eyepiece and is always expressed in mm (millimeters). The range is generally between 2mm and 55mm.

MAGNIFICATION Calculation Examples

1. 8-inch Celestron SCT with a focal length of
   2032mm ÷ 40mm eyepiece = 51x

2. 8-inch Celestron SCT with a focal length of
   2032mm ÷ 20mm eyepiece = 102x

3. 8-inch f/6[1] Newtonian has a focal length of
   1219mm ÷ 20mm eyepiece = 61x

4. 6-inch f/8[1] Newtonian has a focal length of
   1219mm ÷ 20mm eyepiece = 61x

5. 4-inch Tele Vue 101 with a focal length of
   540mm ÷ 10mm eyepiece = 54x

6. 2.75-inch Tele Vue Pronto with a focal length of
   480mm ÷ 10mm eyepiece = 48x

TRUE Field of View

TRUE field of view (field of view is abbreviated FOV) is the diameter, in arc angle[2] degrees, of the circular patch of sky seen through an eyepiece. How much of the sky you see through a particular eyepiece depends on the eyepiece design *and* the magnification that it achieves with a particular telescope.

Knowing the TRUE FOV is useful during celestial exploration of deep sky objects because it 1) allows you to estimate the arc size of objects as well as the arc distance between stars and/or objects, and 2) provides a size reference for matching the eyepiece field of view to the scale of star charts.

To calculate TRUE FOV, you need to know the eyepiece's APPARENT field of view and the magnification that it yields for a telescope. APPARENT FOV is an attribute of the eyepiece design, and, like TRUE FOV, is expressed in arc degrees (°). APPARENT FOVs for popular eyepieces are listed on the next page. Unfortunately, they are not inscribed on eyepieces like focal lengths, but may be indicated on any instructions. Otherwise, check the manufacturer's Internet site. The range of APPARENT FOVs is about 30° to 84°.

---

[1] See Focal Ratio (f/ratio) in the Glossary. [2] The word "angle" is often omitted.

TRUE FOV can be calculated using the following formula:

> APPARENT Field of View of Eyepiece ÷ Magnification from Eyepiece = **TRUE Field of View**

TRUE FOV Calculation Examples

For these examples, the answer is the diameter of the circle, in arc degrees, seen through the eyepiece.

1. A 40mm Plössl eyepiece with a 50° APPARENT FOV yields a magnification of 51x with an 8-inch SCT.
   50° ÷ 51x = 0.98° TRUE FOV

2. A 6mm Nagler Radian eyepiece has a 68° APPARENT FOV and yields a magnification of 90x with a 4-inch Tele Vue 101. **68° ÷ 90x = 0.76°** TRUE FOV

DISCUSSION. The method I discuss here for calculating TRUE FOV is simple and fast, but there are other methods that yield more accurate results but they require more work.

Barlows. If you use a 2x barlow lens with an eyepiece, the APPARENT FOV remains the same but the magnification doubles, so the TRUE FOV is halved. For example, if a 40mm Plössl normally yields 51x, coupled with a 2x barlow it achieves a magnification of 102x, so its TRUE FOV becomes:

$$50° ÷ 102x = 0.49° \text{ TRUE FOV}$$

Changing TRUE FOV from arc degrees to arc minutes. To change results in arc degrees to arc minutes (there are 60 arc minutes in an arc degree), multiply the arc degrees by 60. For examples #1 & #2 above, this yields:

$$0.98° \times 60 = 59' \text{ and } 0.76° \times 60 = 46'$$

APPARENT Field of Views for Common Eyepieces

Plössls: 50°
Orthoscopics: 40° to 45°
Tele Vue **Radians**: 60°
Tele Vue **Panoptics**: 68°
Tele Vue **Naglers** (except the zooms): 82°
Vixen **Lanthanum** Series:
    2mm to 7mm: 45°;   9 mm to 25mm: 50°
Celestron & **Meade** usually list the APPARENT FOVs for their eyepieces in their ads and on their internet sites, but they often do not use the word "APPARENT," and usually just the word "Field."

Views through Eyepieces with Different APPARENT FOVs

Plössl eyepiece achieving 100x has a TRUE FOV of 30'.

Nagler eyepiece achieving 100x has a TRUE FOV of 50'.

These two pictures show the Moon (during a lunar eclipse) at the same magnification, but through eyepieces with different APPARENT FOVs. The Plössl eyepiece shows less area around the Moon because its APPARENT FOV is less than that of a Nagler eyepiece. Different eyepiece designs offer different sized windows for viewing the sky.

How hard is it to manually find Messier's or other deep sky objects? If you can navigate your car using a street map, you should be able to translate the same skills into finding deep sky objects. However, if you have trouble with vehicular navigation, it only means that you might have to practice more. I have listed strategies at the end of this chapter.

Once you gain some experience, the process of manually using a telescope to find Messier objects will become easy. As I was editing this section of the book, I decided to go out and observe for an hour. Using this book's star charts and a 4-inch refractor at 28x, I found 7 Messier star clusters in 7 minutes. I then tried to locate the 5 galaxies below Leo, but could not see them with a 9-day-old Moon in the sky.

Know your telescope. If your telescope is new or you are unfamiliar with using it, I recommend getting acquainted with its operation during the daytime, in a comfortable indoor setting before you go out and observe at night. This will save you from stumbling and becoming frustrated in the dark. The initial star alignments needed for GO TO telescopes can often be simulated during the daytime, so practice the procedure using the hand-controller before you go out to observe.

Telescope accessories
Finderscopes. I highly recommend using or switching to a reflex-sight finder, like the Telrad, for aiming your telescope. Overall, this type of finder is easier and faster to use than traditional finderscopes (which look like little refractors) which have restricted fields of view and magnified imagery that is either upside down or mirror-reversed, all causing confusion when you're trying to match images between the finder, chart and sky.

Accurately align your finder (reflex-sight or traditional) to the view through the eyepiece. Good pointing accuracy of the finder increases the likelihood of finding and relocating objects quickly.

When you point the telescope using your finder, occasionally the deep sky object you are trying to find will be in the eyepiece field of view. However, more often it will not be there and you will have to move the telescope about to locate it — see page 76.

Eyepieces & magnification. Today's eyepieces are great. You can hardly go wrong with any of them. However, I would not go overboard in purchasing eyepieces until you gain some observing experience.

No matter the size of the telescope, *to locate an object, always start with an eyepiece that provides a magnification between 25x and 60x.* Any

Reflex-type finders like the popular Telrad shown below are the best kind of aiming device for pointing telescopes. The Telrad uses a bull's-eye reticle.

Bull's-eye appears to float in the air.

eyepiece that yields low power also provides a wide field of view, which means that it allows you to see more of the sky at one time, making it easier and faster to locate objects. Almost any eyepiece that delivers 50x to 60x magnification has a field of view of at least 1° or more — the equivalent of two Moon widths, which is a nice wide swatch for scooping up Messier's or other deep sky objects.

And most Messier objects look their best with magnifications from 50x to 100x, independent of telescope size. *Higher magnifications are normally not needed or necessary for these objects!*

Other observing considerations when searching for Deep Sky Objects

Personal comfort. Your personal comfort is paramount for an enjoyable observing experience and will be a major factor determining the length of your observing session. Of primary consideration is clothing. Dress appropriately, which usually means to dress warm. This is not a major issue if you are at home, but becomes one at remote locations where you can't run inside for an extra sweater or jacket, so if you observe away, ALWAYS carry additional wear, no matter how warm it might seem when you depart.

Comfortable seating definitely contributes to a more enjoyable observing experience. You may want to invest in a chair, seat or stool that varies in height so you can reach the eyepiece wherever it lands. Don't forget that regular outdoor-type chairs are great for a respite from the scope.

Positioning your telescope. Situate or position your telescope away from bright or glaring lights. If this is not possible, at least keep your back to these lights.

What do deep sky objects look like through a telescope?

Except for some of the brighter open clusters, all DSOs are much, much fainter than photos. Additionally, the vivid colors and miniscule detail in nebulae and galaxies, captured by photos, cannot be seen visually through the eyepiece. Stars and planets show color, but nebulae and galaxies look whitish. Don't let this dissuade you from marching forward to find and observe these objects. Views through the telescope are real and captivating. There is something magical about seeing them "live," a vividness that cannot be conveyed through photographs or even words. Star clusters are very often prettier than in pictures. And when some clusters get close to the horizon, they start to shimmer, providing breathtaking sights. Globular clusters really "come alive" in larger telescopes (like 12-inch and above) and details of objects can be gleaned by spending time studying them. There is a whole universe out there, so go out and have fun seeing it firsthand!

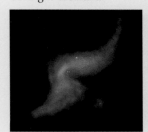 

Comparison of the Orion Nebula as viewed through a small telescope (left) and in a color photo.

Astronomical twilight. Deep sky objects can only be observed when the night is darkest — which is generally between astronomical twilights — starting around 90 minutes after sunset and ending 90 minutes before sunrise.

High in the sky. Like the Planets, Messier objects appear better and brighter when observed higher in the sky away from the more turbulent atmosphere near the horizon.

Night vision. If you try to observe deep sky objects after walking out from a brightly lit interior, you will not be able to see them until your eyes adapt to the dark. This takes several minutes for an initial adaptation (see page 67), to about 15 minutes for a deeper adaptation allowing even fainter objects to be seen.

Red-light flashlight. To preserve your dark-adapted eyes, use a *variable intensity* red-light flashlight to look at your reference material or for writing notes. Using a white-light flashlight interferes more with your night vision than a red light. The ability to vary the light output of a flashlight is very useful for accommodating different levels of dark-adapted eyes. Avoid using flashlights that are too bright, like some red-light LED keychain lights.

Star charts. Although the star charts in this book are good for finding all of the Messier objects, you may eventually want to purchase charts that go to fainter magnitudes, like *Sky Atlas 2000.0* by Wil Tirion and Roger Sinnott (Sky Publishing Corporation). This atlas also indicates many NGC and IC objects for further exploration.

Bright Seasonal Picks

SPRING M44 cluster*, M81/82 galaxy pair, M3 globular.

SUMMER M7 cluster*, M6 cluster*, M22 globular, M8 nebula, M13 globular, M92 globular, M17 nebula, M10/M12 globulars, M27 planetary nebula.

FALL M31*/M32 galaxies, M111/M112 clusters, M34 cluster.

WINTER M45 cluster*, M42/M43 nebula*, M41 cluster, M35/36/37/38 clusters.

*Very easy binocular objects.

Additional advice for first time deep sky observers

Start with the brightest objects. If the thought of finding deep sky objects seems daunting, start with the biggest and brightest objects first, easing into the "harder" ones later. To the left is a list of some of the easiest Messier objects to find — most are visible in light polluted skies.

Practice strategies. Initially, I recommend that your observing sessions for finding Messier objects be short periods of time — 30 minutes or so — instead of long, drawn-out sessions, but if you are on a roll, don't stop. If finding these objects is difficult for you, plan on many short sessions over a period of time and stick with the easiest objects noted in the catalogue pages. You *will* improve with practice. If you are having trouble identifying objects, your ability to recognize them *will* get better as you view more, and gain experience.

Seeking help. If you are having or continue to have difficulty finding Messier objects or are uncertain about the whole process, ask an interested friend for help (two minds are often better than one), join an astronomy club, or get advice from a local telescope store. Spending an hour with a more experienced amateur will get you going the fastest. You can then repay the favor by helping someone else get started. If you are really at wits' end and are serious about the pursuit of this wonderful hobby, email me for help at ken@kenpress.com.

You can do it! If you walk around your neighborhood and then find your way home, you can learn to use a telescope and find deep sky objects!

Strategies for manually finding Messier or other Deep Sky Objects

STRATEGY 1
Plain as the nose on your face.
Several Messier objects can be seen by most observers with the naked eye, even in light-polluted skies. These include M42, M45 and M31. Other objects like M44, M111/M112 and M7 can be seen with the naked eye under darker skies. Naked-eye objects are the easiest to locate and view through a telescope because all you have to do is point the telescope at them using your finder. Observers vary in their ability to see objects with the naked eye. Some

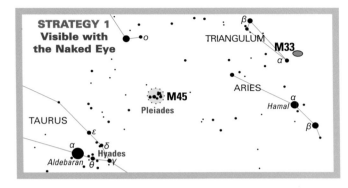

of my friends can see the M33 galaxy in dark skies while I can barely see it using my telescope.

STRATEGY 2
One eyepiece field of view away.
Most Messier objects require a telescope to see, which means that some work is involved in finding them. Given this, there are several objects that are within one or a few 1° eyepiece fields of view from naked-eye stars (remember, any modern eyepiece that yields a magnification of around 50x provides at least a 1° field of view). So, finding these objects is simplified because all you have to do is point the telescope at these stars and move the scope a bit to find the object. Examples are as follows:

M77 is within 1° of δ *Ceti*[1] whereas M74 is $1\frac{1}{3}$° from η *Piscium* (shown on the next page).

M109 is less than 1° from *Phad* while M108 is $1\frac{1}{2}$° and M97, $2\frac{1}{3}$° from *Merak*, stars that make up the Big Dipper bowl (see Chart K). M1 is about 1° from the horn star ζ in Taurus (see Chart H).

---

[1] See page 80 and the bottom of page 336 for information about this nomenclature.

Using Simple Geometric Figures to Manually Find Messier Objects

How do you locate the "harder to find" Messier objects manually when they aren't near any conspicuously bright stars? A common navigational strategy is to point your finder "at" the object, using its position to form a line or triangle with other nearby naked-eye stars.

For example, several Messier and other deep sky objects are located on lines extending from two visible stars. Other deep sky objects often form the "invisible" *vertex* (the "point" where two lines join in a triangle or other geometric figure) that completes a triangle with two other naked-eye stars. So, using lines and creating triangles with naked-eye stars and your target object provides a spot or starting point in the sky to point your finder at for locating objects.

Shown below are common geometric shapes which an "invisible" deep sky object often forms with naked-eye stars. Personally, I use the *line segment* most often, followed by the *"shallow" obtuse isosceles triangle*.

You must decide, based on the placement of naked-eye stars, which geometric shape to use for aiming your telescope.

Once you aim your telescope, using the finder, at the approximate location of your target object, you will then have to nudge or slightly move the telescope about — up and down and left to right — until you come across the object. If you move too far away from the target spot (which is easy to do), move the scope back to it using the finder, and search again. I often have to return to my original target spot several times and hunt again before I find an object.

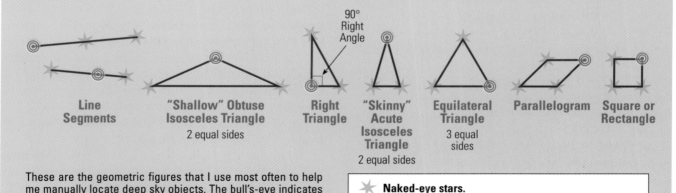

These are the geometric figures that I use most often to help me manually locate deep sky objects. The bull's-eye indicates the position of the "invisible" object which is the spot to point the finder at. *You* must decide on the geometric shape, and thus the location of the bull's-eye, using a star chart.

⭐ **Naked-eye stars.**

◎ **Bull's-eye. Where the Finder is pointed — a starless spot at the approximate location of a Messier object.**

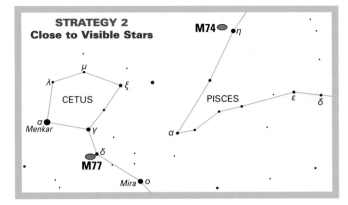

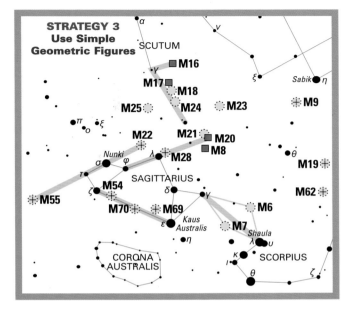

Although M77 and M74 are easy to locate, these galaxies are two of the fainter Messier objects. Seeing them will be much harder if you don't have good dark skies. The same can be said about M108 and M97.

STRATEGY 3
Oh no, it's done with geometry!

I have found that many people have a distaste for geometry. I have good news and bad news. The bad news is that working with some geometry is essential to manually finding deep sky objects. The good news is that it involves the simplest form of geometry, which means that everyone can do it.

So, how do you locate faint objects in the sky that appear to lie in the middle of nowhere? Well, with the use of star charts like those in this book, I and many others locate these objects using simple geometric figures in combination with surrounding naked-eye stars.

So, when a Messier object is not plainly visible to the naked eye or near a naked-eye star, one can often locate it by "using" a little geometry. Frequently deep sky objects fall on a line formed by two naked-eye stars, as is the case with M55, M70 and others shown above. Now, if this is not the case, objects often create shallow triangles with visible stars, as happens with M6 and M7, also shown above. In many instances, using a "geometric" strategy provides a good starting point for aiming the finder to locate an object. Remember, once the telescope is aimed, you will most likely have to move it around to "bump" into the object (see page 76).

Here are some comments about the objects indicated in the above chart.

1. M55 is on a line formed by the stars τ *Sagittarii* and *Nunki* and twice this length away from τ.

2. M70 is halfway between ζ and ε *Sagittarii*. M69 is above this line on the west end and M54 is just above this line, but more toward ζ on the east end.
3. M28 is about 1 degree away from λ on the φ to λ line.
4. M22 is on a line formed by the stars τ *Sagittarii* and *Nunki* at 1½ times this length from *Nunki*. Another way to look at this is that M22 forms the vertex of a "rough" parallelogram with *Nunki*, φ and λ *Sagittarii*.
5. M6 and M7 form shallow obtuse isosceles triangles with γ *Sagittarii* and *Shaula*.
6. M8, M20 and M21 are just a little over a one length extension of the line drawn between φ and λ *Sagittarii*.

STRATEGY 4
Star hopping with a star chart.

Jumping from star to star, called, star hopping, using a star chart as your guide, is a sure-fire way to move among the stars. You navigate by guiding the telescope through the eyepiece, referring to a chart for directions. This is not a difficult task, just very tedious because of the time it takes to match up and verify the stars in the eyepiece with those on the chart. This is why star hopping is usually used as a last resort to find objects.

To star hop, you need detailed star charts like Charts U and V, or a fainter magnitude atlas that covers the whole sky, like *Sky Atlas 2000.0* (Sky Publishing Corporation).

How do you begin the process of star hopping to a target object? First, point the telescope to a naked-

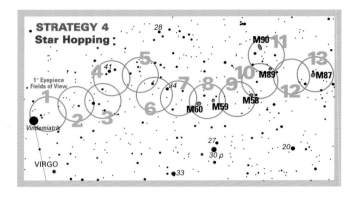

eye star, like *Vindemiatrix*, shown above, using an eyepiece providing a liberal field of view, say around 1° or so (*remember*, any eyepiece yielding 50x will provide about a 1° field of view). *Stay with lower magnifications so you have plenty of stars in the eyepiece to match with those on the chart!* Orientate the chart to the stars in the eyepiece. Now, ONLY move the scope ONE eyepiece field of view at a time, in the direction you want to go, matching the stars in the eyepiece with those on the chart, until you get to your object. Although this is slow and tedious — but sure-fire — pay attention because if you lose track of where you are, you'll have to start over.

Note for refractor and SCT users. If you are using a 90° diagonal for comfortable viewing, the image through the eyepiece is a mirror image, meaning that left and right are reversed. This will make correlating with stars on "correctly" printed charts a little tricky. One way around this is to scan the chart into your computer, "flip it" in a graphics program, then print it out. Also, planetarium software like *The Sky*, can automatically print out mirror-reversed star charts.

Using the Star Charts

Twenty-two charts are provided on the following pages to help you manually find the Messier objects. These include two summary charts showing every Messier object as well as detailed charts to aid in the positive identification of the galaxies in the Virgo Galaxy Cluster. There is also a mirror-reversed Virgo Galaxy Cluster chart for those using refractors or SCTs with diagonals.

In this book, I assume readers possess the skill to identify the constellations. One of the best devices for orientating oneself to the night sky is a planisphere, since these circular charts can be adjusted to any time and day of the year. Once you know which constellations are out, you can then choose the most appropriate charts in this book. To help get you started, the table on page 81 indicates the approximate times when a chart is centered on your local meridian.

Elements of the star charts are discussed below, and examples referencing the letters are on the next page.

A **Constellation names**. The names of constellations are in UPPERCASE letters. A list of the constellations, their abbreviations and genitive form are in the Appendices.

B **Star names**. The names of stars throughout this book are in *italics*. Their spelling was taken from the *Millennium Star Atlas* published by Sky Publishing Corporation and the European Space Agency. The spelling of star names differs among charts and atlases because many of the names originate from Arabic, which has an alphabet consisting of non-roman characters. So the English spelling is based on oral transliteration, which varies, depending on the translator.

C **Blue constellation lines**. There is no universal agreement on the use and placement of constellation lines because they have no scientific value. Today's lines generally follow major features or parts of figures drawn on ancient charts. One of the earliest star atlases was in the *Almagest* by Ptolemy, circa 150 A.D. which describes the position of stars in relation to the celestial figures. An example of an early chart is shown on page 45.

D **Asterisms**. The names of several asterisms, that is, parts of constellations, are indicated in **bold**, like the "**Belt**" of ORION or "**Circlet**" of PISCES.

E **Right Ascension lines**. Analogous to lines of longitude on the Earth, but there are only 24 major divisions for the stars, numbered 0 hours to 24 hours and representing the rotation of the celestial sphere. These 24 hours of division are *4 minutes* less than our clock-time and are called sidereal time — see *Sidereal Day* in the Glossary.

F **Declination lines**. Analogous to lines of latitude on Earth and use a similar nomenclature. The celestial equator is 0°. Declinations north of the celestial equator are indicated by a plus sign, up to

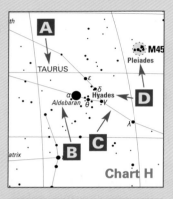

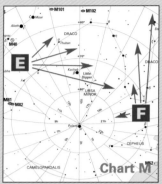

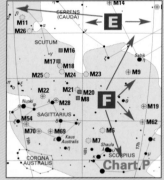

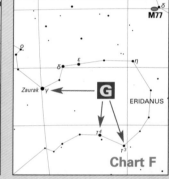

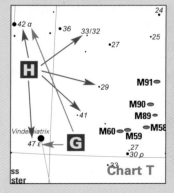

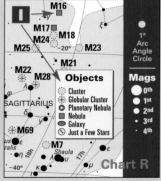

+90° for the North Celestial Pole. Declinations south of the celestial equator are indicated by a negative sign up to −90° for the South Celestial Pole. A Declination degree is equal to one arc angle degree.

G **Greek Bayer letters.** Next to many of the brighter stars is a lowercase Greek letter, a nomenclature originated by Johann Bayer (1572–1625). The lettering starts anew with each constellation. Generally, Greek letters were assigned to the stars by order of magnitude. Occasionally, nearby stars are designated with the same letter but differentiated by superscripts. Bayer designated stars with a Roman letter if he had exhausted the Greek alphabet, a practice since abandoned but used in Messier's time and indicated in his descriptions.

To refer to a star using its **Bayer** letter or **Flamsteed** number, see the examples at the bottom of page 336.

H **Flamsteed numbers.** John Flamsteed (1646–1719), appointed the first Astronomer Royal of England, designated stars by numbering them. Like the Bayer letters, his numbering starts anew with each constellation but is ordered by Right Ascension instead of magnitude. Every Bayer lettered star also has a Flamsteed number. In this book, Flamsteed numbers are used on the detailed charts and for other pertinent stars without Bayer letters.

I **Object symbols.** A total of 6 symbols are used to distinguish objects. There are a few objects, like M16, M17 and M20 whose symbols could be disputed because they are a combination of two types of objects — nebulae and clusters. And there is the case of M24, which Messier describes as a cluster but represents a thick part of the Milky Way. In these instances, I chose the symbols by which these objects are commonly recognized today.

Guide for Chart Usage

The *center* Right Ascension of the following charts will be on your local meridian at approximately the times indicated below (Standard Time).

	Around 8 p.m.	Around Midnight	Around 4 a.m.
January	E/F	G/H	K/L
February	G/H	I/J	K/L
March	G/H	K/L	M/N
April	I/J	K/L	O/P
May	K/L	M/N	O/P
June	K/L	O/P	Q/R
July	M/N	O/P	C/D
August	O/P	Q/R	C/D
September	O/P	C/D	E/F
October	Q/R	C/D	G/H
November	C/D	E/F	G/H
December	C/D	G/H	I/J

NOTE: Add one hour to these times for Daylight Saving Time from April to October.

Chart Scales

Chart	Scale in Arc Degrees	½° & 1° Eyepiece Fields of View
A&B	 1° 5° 10° 20° 30° — 6 mm or ¼ inch per 5 arc degrees	
C-R	 1° 5° 10° 20° — 2½ mm or ¹⁄₁₀ inch per arc degree	
S&T	 1° 5° — 6 mm or ¼ inch per arc degree	
U&V	 1° 5° — 12 mm or ½ inch per arc degree	

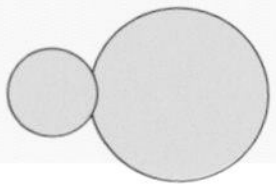

Greek Alphabet

	Upper Case	Lower Case & Variations*
Alpha	A	α
Beta	B	β
Gamma	Γ	γ, γ
Delta	Δ	δ
Epsilon	E	ε
Zeta	Z	ζ, ζ
Eta	H	η
Theta	Θ	θ, ϑ
Iota	I	ι, ι
Kappa	K	κ
Lambda	Λ	λ
Mu	M	μ
Nu	N	ν
Xi	Ξ	ξ, ξ
Omicron	O	o
Pi	Π	π, ϖ
Rho	P	ρ
Sigma	Σ	σ
Tau	T	τ, τ
Upsilon	Y	υ, υ
Phi	Φ	φ, φ
Chi	X	χ, X
Psi	Ψ	ψ
Omega	Ω	ω

*Only the lowercase letters of the Greek alphabet are used for designating the brightest stars in each constellation.

81

ARIES

2h

M33

Mirach

ANDROMEDA

M31

M74

M32 M110

PISCES

Alpheratz

0h

Great
Square

PEGASUS

Scheat

LACERTA

EQUULEUS

DELPHINUS

M76

M103

CASSIOPEIA

Caph

M52 CEPHEUS

Milky
Way

4h 6h

Polaris

North Star

Little
Dipper

Kochab

URSA
MINOR

8h

LYNX

M82 M81

URSA MAJOR

Dubhe

Merak M108

Big Dipper M97

Megrez Phad

M40 M109

Alioth

Mizar

M101

M102 Alkaid

M39

Deneb

CYGNUS

M29

Vega

LYRA

M56

M57

Albireo

M27

DRACO

Thuban

Etamin

+60°

M92

M13

Keystone

LEO
MINOR

CAN

M

Sic

M106

M51

M63 CANES
VENATICI

M94

COMA
BERENICES

M3 M64 M85 M98

BOOTES

HERCULES

CANC

M100

M91 M90 M84

M53 M90 M87

M89 M49

M60

M59 M58

Den

M99
M86

M27

VULPECULA

M15

M71

Enif

SAGITTA

M2

M73 M72

AQUARIUS

CAPRICORNUS

Altair

AQUILA

M30

Mags

0th

1st

2nd

3rd

PISCIS
AUSTRINUS

GRUS

Alnair

M75

22h

20h

MICROSCOPIUM

CORONA
BOREALIS Alphekka

Rasalhague

SERPENS
(CAPUT)

Kornephoros

M5

OPHIUCHUS

0°

M11

SERPENS
(CAUDA)

M26

SCUTUM

M16

M17

M18

M25

M24

M22

Nunki

SAGITTARIUS

M55

M54 M28

M70 M69

Kaus
Australis

M7

M14

M10 M12

18h

16h

M107

Sabik

M23

M21

M20

M8

M9

M19 M4

Antares

M62

Shaula

SCORPIUS

M6

−30°

Arcturus

VIRGO

Spica

M61

M104

CORVUS

LIBRA

HYDRA M68

M80

M83

LUPUS

Menkent

CENTAURUS

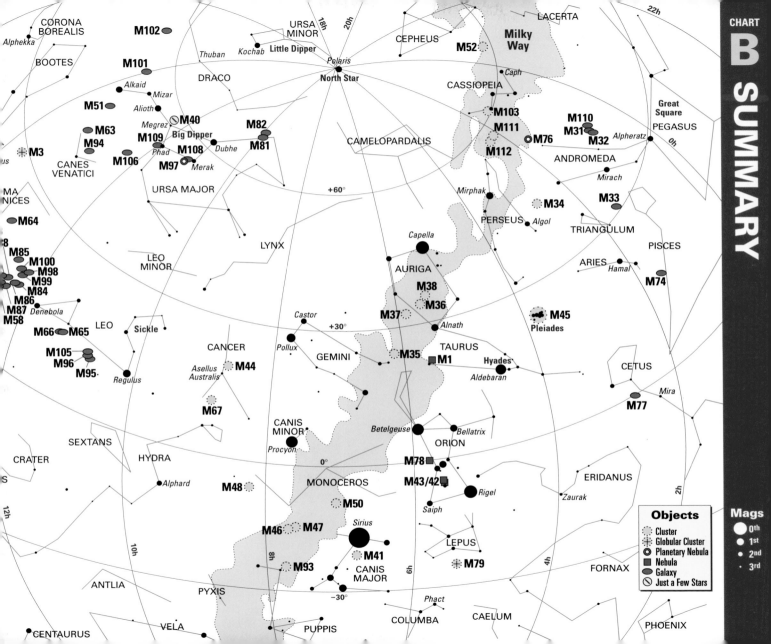

CORONA BOREALIS
Alphekka
BOOTES

M102

URSA MINOR
Thuban
Kochab **Little Dipper**
Polaris
North Star

18h
20h

CEPHEUS
M52
LACERTA

Milky Way

22h

CASSIOPEIA
Caph

M101
Alkaid
Mizar
DRACO
Alioth

M51

M63
M94
M3

M109
Megrez **M40**
Big Dipper
Phad **M108**
Dubhe

M82
M81

CANES VENATICI
M106
M97
Merak

URSA MAJOR

CAMELOPARDALIS

+60°

M103
M111
M112
M76

M110
M31
M32
Alpheratz

Great Square
PEGASUS
0h

ANDROMEDA
Mirach

COMA BERENICES
M64

LEO MINOR
LYNX

Mirphak
PERSEUS
Algol
M34

M33
TRIANGULUM

PISCES
ARIES
Hamal

M74

M85
M100
M98
M99
M84
M86
M87
M58
Denebola

LEO
Sickle

Castor
+30°
Pollux
GEMINI

M38
M36
M37
Alnath

Capella
AURIGA

M45
Pleiades

CETUS

M66
M65

M105
M96
M95
Regulus

CANCER
Asellus Australis
M44

M35
TAURUS
M1
Aldebaran
Hyades

Mira

M77

SEXTANS

CANIS MINOR
Procyon

CANIS MINOR

Betelgeuse
ORION
Bellatrix

0°

CRATER
HYDRA
Alphard

M48
MONOCEROS

M50

M78
M43/42
Saiph
Rigel

ERIDANUS
Zaurak

12h
10h

M46
M47
Sirius

8h

M41

LEPUS
M79

FORNAX
2h

M93
CANIS MAJOR

-30°
6h

Phact

4h

ANTLIA
VELA
PYXIS
PUPPIS

COLUMBA
CAELUM

PHOENIX
CENTAURUS

Objects
⊙ Cluster
✳ Globular Cluster
⊙ Planetary Nebula
■ Nebula
⬭ Galaxy
⦸ Just a Few Stars

Mags
○ 0th
● 1st
• 2nd
· 3rd

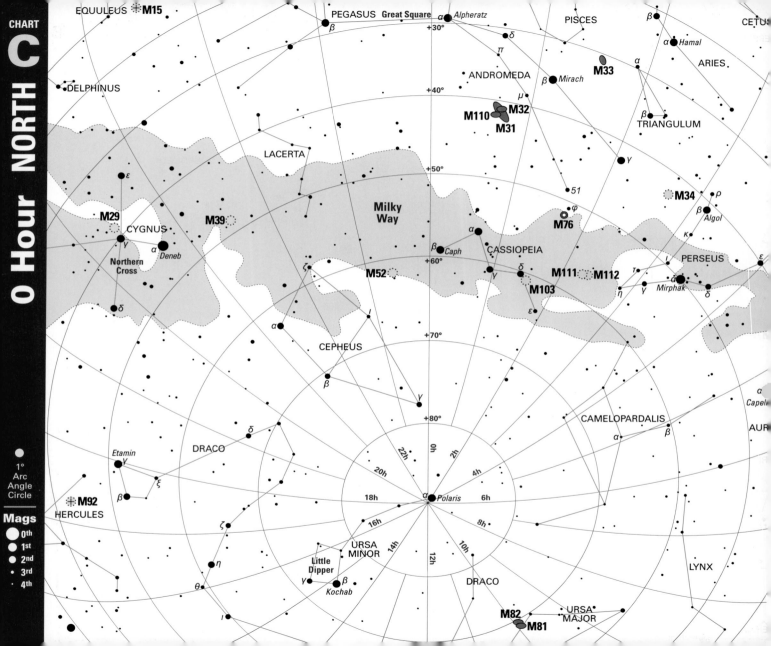

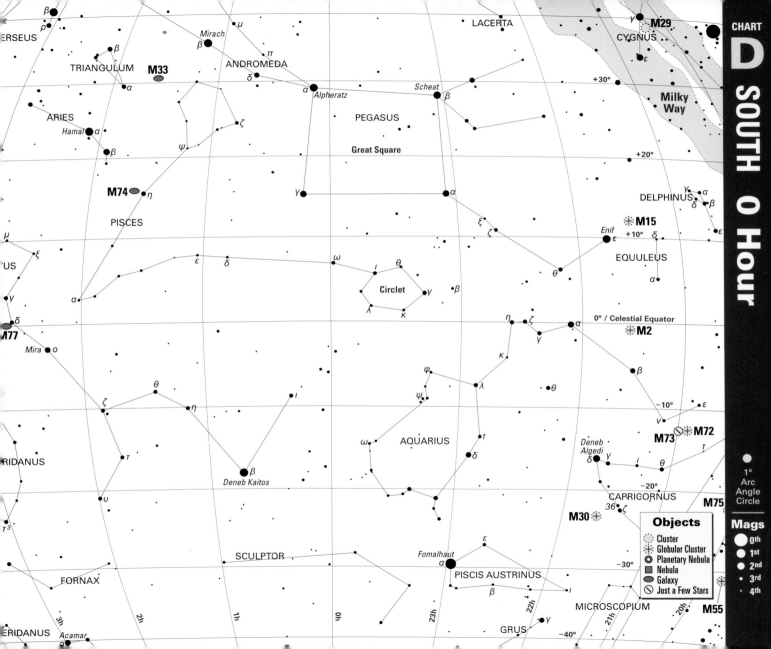

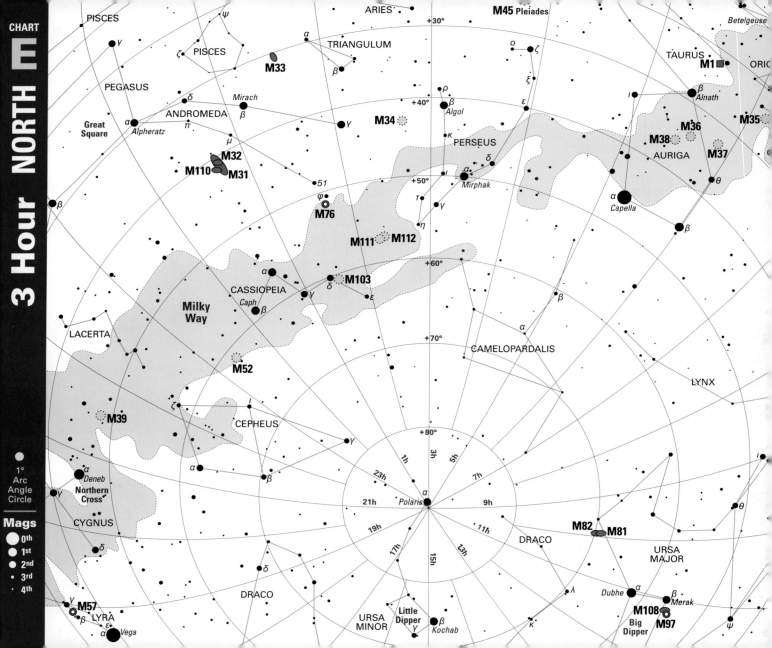

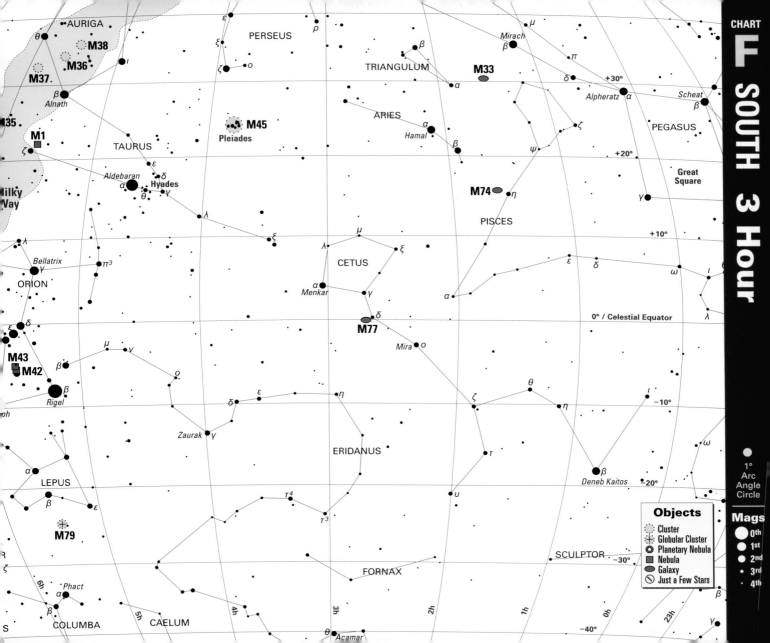

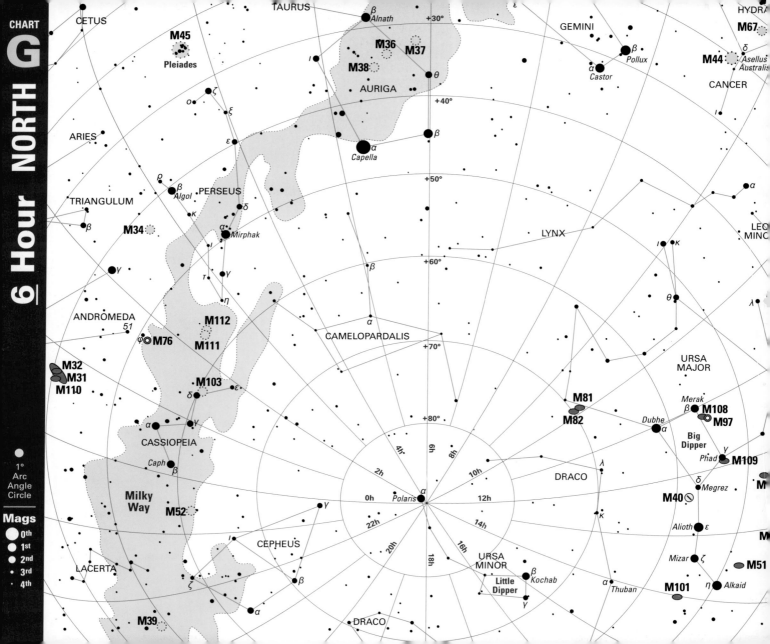

CHART

G

6 Hour NORTH

1°
Arc
Angle
Circle

Mags

- 0th
- 1st
- 2nd
- 3rd
- 4th

CETUS

TAURUS

β
Alnath

+30°

GEMINI

HYDRA

M67

M45
Pleiades

M36
M38
M37

θ

β
Pollux

M44

*Asellus
Australis*

δ

AURIGA

α
Castor

CANCER

+40°

ARIES

ο

ζ

ξ

ε

β

α
Capella

+50°

ρ

β
Algol

PERSEUS

LYNX

LEO
MINC

TRIANGULUM

β

M34

δ

κ

+60°

α

LEO
MINC

α
Mirphak

β

ι
κ

ANDROMEDA

ι

γ

η

α

θ

λ

51

τ

CAMELOPARDALIS

+70°

URSA
MAJOR

φ

M112

M76

M111

M32

M31

M110

M103

δ

ε

+80°

M81

M82

Merak

β

M108

M97

Dubhe

α

**Big
Dipper**

α

γ

Phad

M109

CASSIOPEIA

γ

4h

6h

8h

DRACO

Caph

β

2h

10h

λ

δ

Megrez

M40

**Milky
Way**

M52

0h

Polaris

α

12h

M

22h

14h

Alioth

ε

CEPHEUS

γ

20h

16h

URSA
MINOR

Mizar

ζ

M51

ι

18h

β

Kochab

η

Alkaid

LACERTA

ζ

**Little
Dipper**

α

Thuban

M101

β

γ

M39

DRACO

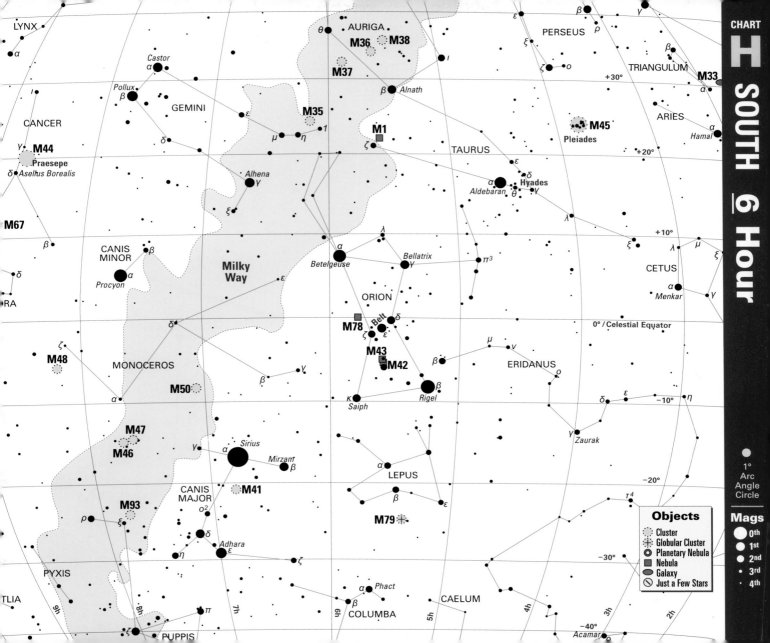

LYNX
α

AURIGA
θ
M36 M38
M37

PERSEUS
ε
ρ
ξ
ζ o
γ

TRIANGULUM
β
M33
α

Castor
α
Pollux
β
GEMINI

ε
M35
μ η
ι

β Alnath

+30°

ARIES
α
Hamal

CANCER

γ M44
Praesepe
δ Asellus Borealis

δ

Alhena
γ

ξ

TAURUS

M45
Pleiades

ε δ
α Hyades
Aldebaran θ γ

+20°

M67

β

CANIS
MINOR

β

λ

ξ
λ μ
ξ

+10°

RA

δ

Procyon
α

Milky
Way

ε

α
Betelgeuse

λ

Bellatrix
γ

π³

CETUS

α
Menkar
γ

M48

ζ

MONOCEROS

β γ

ORION
Belt
ζ δ
ε

0° / Celestial Equator

M78

μ ν

ERIDANUS

M50

δ

α

M43
M42

β

o

M47

β

M46

γ

Sirius
α
Mirzam
β

κ
Saiph

Rigel
β

γ Zaurak

δ ε
−10° η

CANIS
MAJOR

M41

α
LEPUS
β ε

τ⁴

−20°

M93

ρ

ξ

o²

δ

Adhara
η ε

ζ

M79

−30°

PYXIS

α Phact
β
CAELUM

Acamar
−40°

9h

8h 7h

6h 5h 4h 3h 2h

COLUMBA

PUPPIS

1°
Arc
Angle
Circle

Mags

0th
1st
2nd
3rd
4th

Objects

⬤ Cluster
✳ Globular Cluster
◉ Planetary Nebula
◼ Nebula
⬭ Galaxy
◎ Just a Few Stars

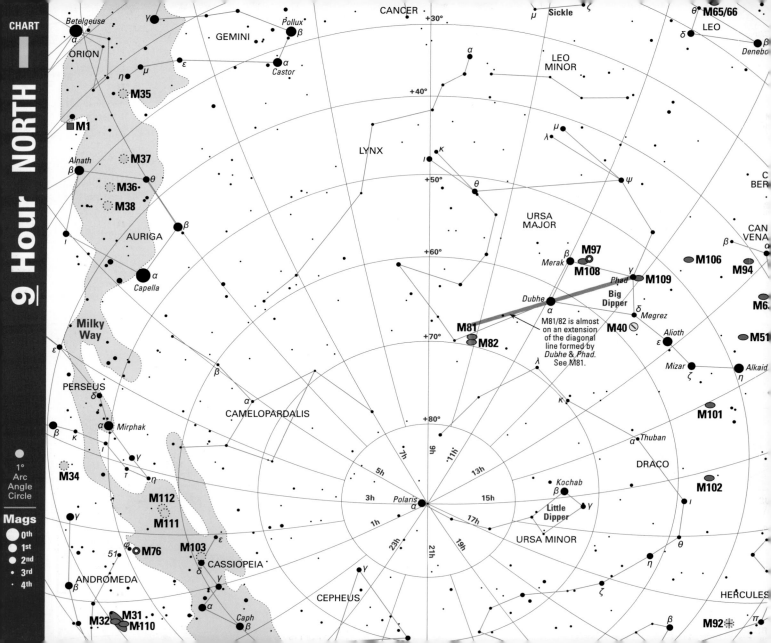

CHART
1

9 Hour NORTH

ORION
Betelgeuse
α
γ
GEMINI
Pollux
β
Castor
α
ε
η
μ
M35
M1
CANCER
+30°
Sickle
ζ
μ
θ **M65/66**
LEO
δ
β
Denebo
LEO MINOR
α
LYNX
+40°
κ
ι
θ
μ
λ
ψ
C BER
CAN VENA
β
α
Alnath
β
M37
θ
M36
M38
AURIGA
ι
+50°
URSA MAJOR
M106
M94
Capella
α
+60°
β
Merak
M97
M108
γ
Phad
M109
Big Dipper
M6
Milky Way
ε
β
Dubhe
α
δ Megrez
M40
Alioth
ε
M51
+70°
M81
M82
M81/82 is almost on an extension of the diagonal line formed by *Dubhe & Phad.* See M81.
Mizar
ζ
η
Alkaid
λ
PERSEUS
δ
+80°
CAMELOPARDALIS
α
α
κ
M101
α
Mirphak
β
κ
ι
γ
τ
η
M34
Thuban
α
DRACO
M102
1° Arc Angle Circle

Mags
● 0th
● 1st
● 2nd
• 3rd
• 4th

M112
M111
φ
51
M76
M103
ε
δ
γ
CASSIOPEIA
7h
9h
11h
5h
13h
3h
Polaris
α
15h
Kochab
β
γ
Little Dipper
ANDROMEDA
β
1h
23h
21h
19h
17h
URSA MINOR
θ
M31
M32
M110
φ
Caph
β
CEPHEUS
γ
η
ζ
HERCULES
β
π
M92

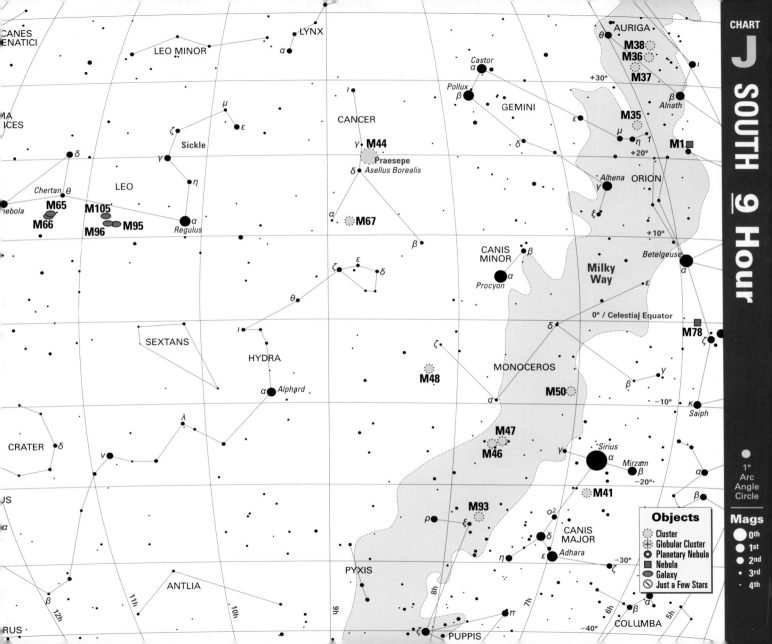

CANES
VENATICI

LYNX

LEO MINOR

α

AURIGA
θ
M38
M36
+30°
M37
β
Alnath
ι

Castor
α
Pollux
β
GEMINI
M35
ε
μ η ι
δ
+20°
M1

CANCER

ι

μ
ζ
ε
Sickle
γ
η
δ
MA
ICES

δ

LEO

Chertan θ
M65
nebola
M66
M105
M96 M95
α
Regulus

γ
Praesepe
δ Asellus Borealis

α
M67
β

M44

Alhena
γ
ORION
+10°
ξ

CANIS
MINOR
β
α
Procyon

Betelgeuse
α

ε

Milky
Way

ζ ε
δ
θ

M78
ζ

0° / Celestial Equator

δ

SEXTANS

HYDRA

ι

MONOCEROS
ζ

M48

M50
γ
β
−10°
κ
Saiph

CRATER δ

ν
λ

α Alphard

α

M47
M46

Sirius
γ α
Mirzam
β
−20°

1°
Arc
Angle
Circle

RUS
α

M93
ρ
ξ

M41

PYXIS
ANTLIA

o²
δ
CANIS
MAJOR
Adhara
η ε
ζ
−30°

α
β

β

Mags
0th

1st
2nd
3rd
4th

β
12h
11h
10h
9h

8h
7h

6h
5h

α
COLUMBA
−40°

π
ζ
PUPPIS

Objects
◌ Cluster
⊕ Globular Cluster
◉ Planetary Nebula
■ Nebula
⬭ Galaxy
⊘ Just a Few Stars

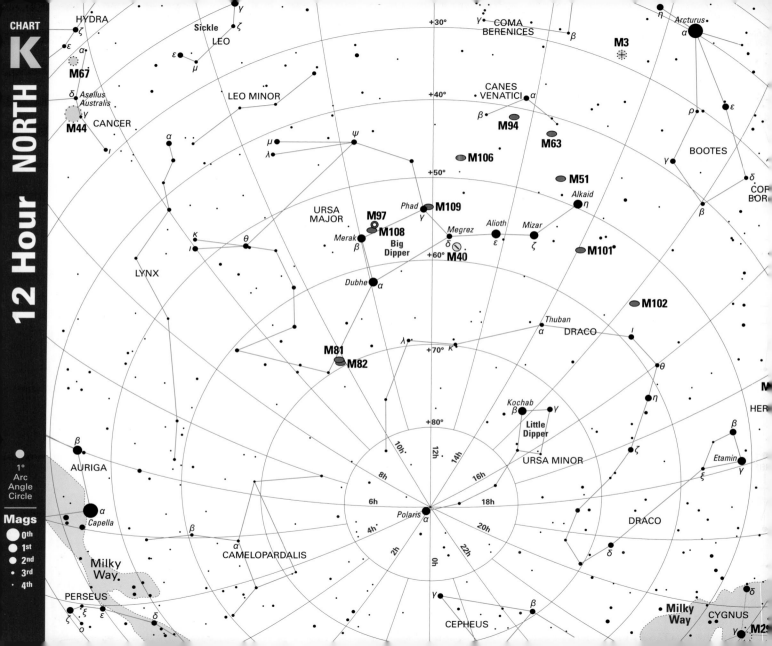

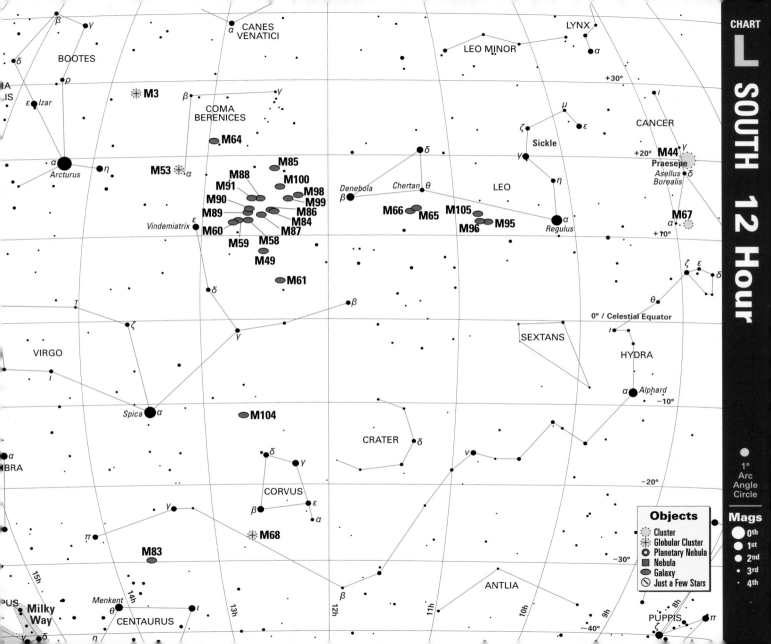

CHART

L

SOUTH 12 Hour

BOOTES

β

γ

δ

ε Izar

ρ

IA
IS

CANES
VENATICI

✳ **M3**

COMA
BERENICES

β

γ

M64

α

Arcturus

η

M53 ✳ α

M88

M91

M90

M89

M60

Vindemiatrix ε

M85

M100

M98

M99

M86

M84

M87

M59 **M58**

M49

M61

δ

LYNX

LEO MINOR

α

μ

ζ

ε

Sickle

γ

η

CANCER

+30°

+20° **M44**
Praesepe

Asellus
Borealis

γ

δ

M67

α

+10°

δ

LEO

Denebola
β

Chertan θ

M66

M65

M105

M96 **M95**

α

Regulus

Spica α

τ

ζ

γ

β

γ

VIRGO

ι

M104

θ

ι

SEXTANS

HYDRA

α Alphard

−10°

0° / Celestial Equator

ζ ε

δ

−20°

CRATER

δ

δ

γ

M83

γ

β

ε

α

CORVUS

ν

✳ **M68**

π

14h

15h

Menkent
θ

CENTAURUS

ι

13h

12h

β

11h

10h

ANTLIA

9h

8h

ζ

PUPPIS

π

−30°

−40°

Milky
Way

α
δ

BRA

α

Objects

⊛ Cluster
✳ Globular Cluster
◉ Planetary Nebula
▪ Nebula
⬭ Galaxy
⊘ Just a Few Stars

1°
Arc
Angle
Circle

Mags

○ 0th
○ 1st
○ 2nd
• 3rd
· 4th

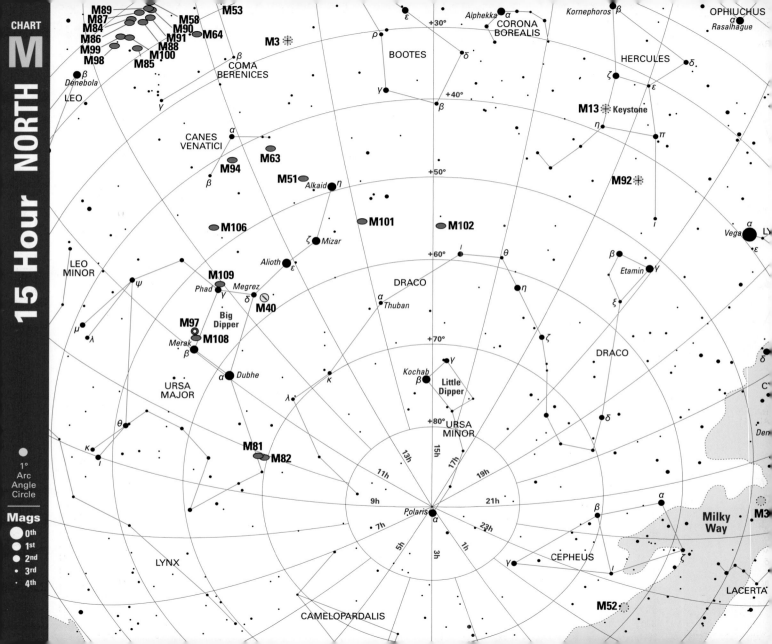

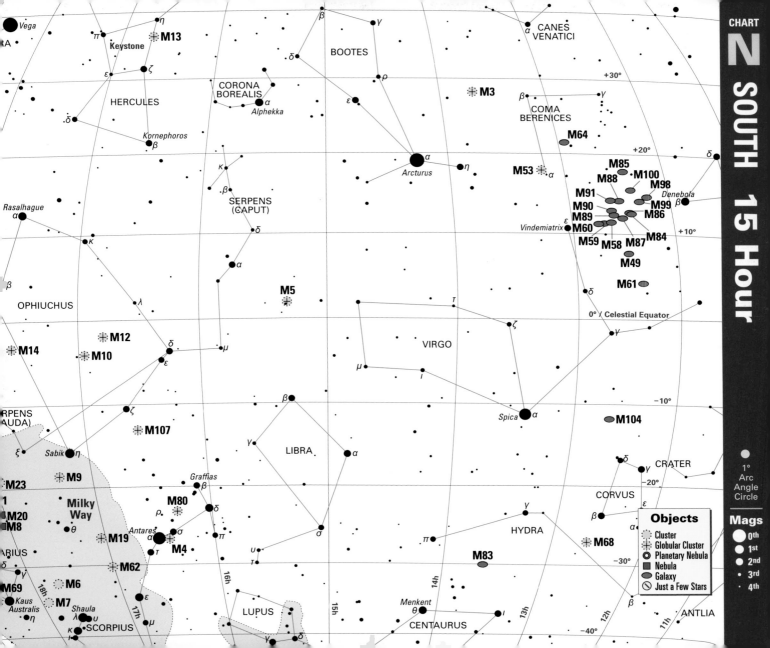

Vega

π η **M13**

Keystone

HERCULES

ε ζ

δ

Kornephoros

β

Rasalhague

α

κ

β

OPHIUCHUS

M14

M12

M10

δ

ε

ζ

M107

ξ

Sabik η

RPENS
(AUDA)

M23

M9

Milky
Way

M20

M8

M19

θ

Antares

α

σ

M4

τ

ρ

π

M62

M6

ARIUS

δ

γ

M69

M7

Kaus
Australis

Shaula

λ υ

η μ

κ **SCORPIUS**

β

BOOTES

γ

δ

ρ

ε

CORONA
BOREALIS

α

Alphekka

κ

β

SERPENS
(CAPUT)

δ

α

M5

μ

β

γ

α

LIBRA

σ

υ

τ

M80

δ

σ

π

16h

15h

LUPUS

γ δ

CANES
VENATICI

α

M3

COMA
BERENICES

β γ

M64

M53 α

M85

M88 **M100**

M91 **M98**

M90 **M99**

M89 **M86**

Vindemiatrix ε **M60**

M59 **M58** **M87** **M84**

M49

M61

δ

Arcturus α

η

+30°

+20°

+10°

0° Celestial Equator

Denebola

β

VIRGO

τ

ζ

μ

ι

Spica α

M104

–10°

δ

γ

CRATER

δ

γ

ε

CORVUS

β

α

HYDRA

γ

π

M68

–20°

–30°

M83

14h

13h

Menkent

θ

ι

12h

β

11h

ANTLIA

–40°

CENTAURUS

δ

Objects

⊛ Cluster
⊕ Globular Cluster
⊙ Planetary Nebula
■ Nebula
⬭ Galaxy
⊘ Just a Few Stars

1°
Arc
Angle
Circle

Mags

● 0th
● 1st
○ 2nd
• 3rd
· 4th

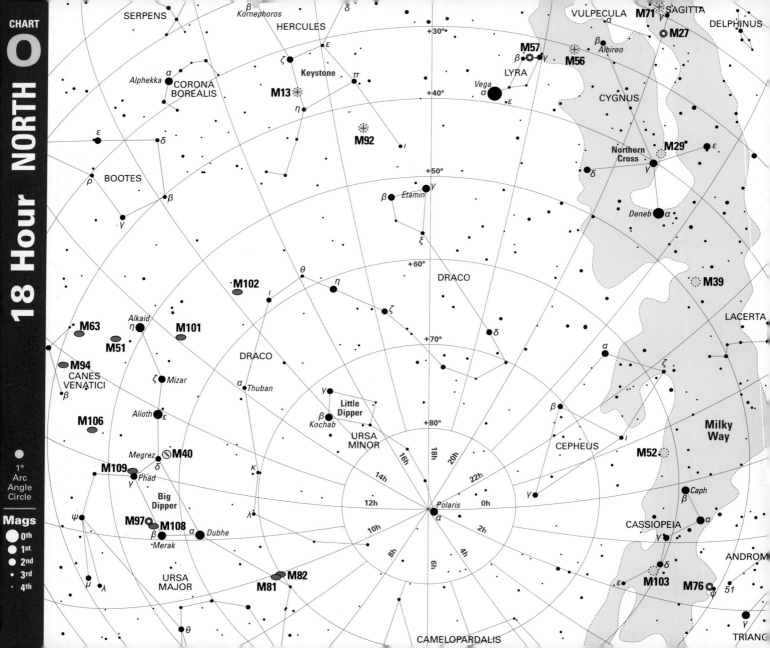

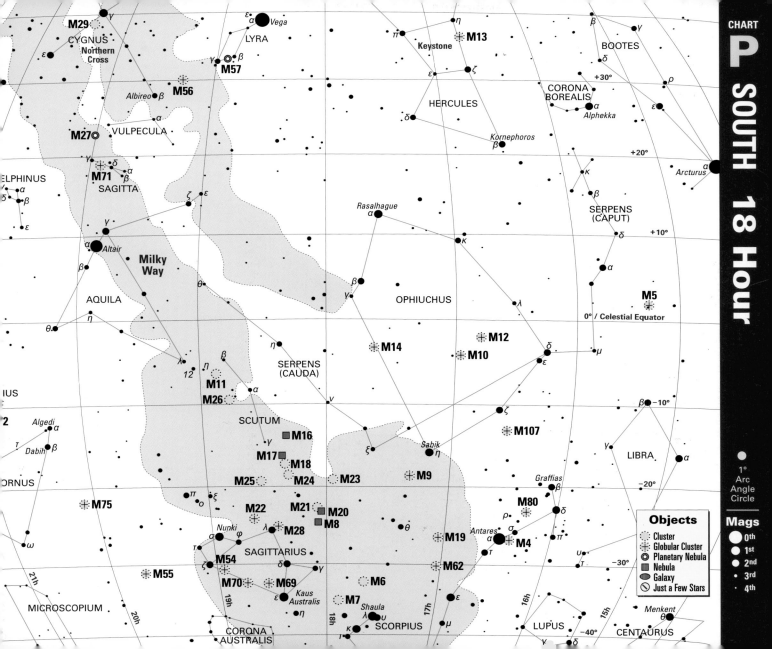

M29
CYGNUS
Northern Cross
ε
γ

Vega
ε
α
LYRA

η
Keystone
M13

β
γ
BOÖTES
δ

Albireo β
M56

γ β
M57

ε
ζ
HERCULES
α
Alphekka
CORONA BOREALIS
ρ
+30°

VULPECULA
M27

δ
α

ε
+20°
Arcturus α

ELPHINUS
α
δ β
ε

γ δ α
β
M71
SAGITTA

Kornephoros
β

κ
β
SERPENS (CAPUT)
δ
+10°

γ
ζ ε
Altair
α

Rasalhague
α

κ

0° / Celestial Equator

Milky Way

β
γ
OPHIUCHUS

λ
α
M5

AQUILA
η
θ

η
θ

M14
M12
M10
δ
μ

λ
η
β
12
M11
M26

SERPENS (CAUDA)

ν

β −10°

IUS
2
Algedi α
τ
Dabih β

SCUTUM

M16
γ

M107
Sabik
η
ξ

LIBRA
γ
α

ORNUS

M17
M18
M24
M25

M23
M9
Graffias β
−20°

1° Arc Angle Circle

M75
π
ο ξ

M22
M21
M8
M20

θ

M80
δ
ρ
σ
π

Objects
⊛ Cluster
✳ Globular Cluster
⊙ Planetary Nebula
▪ Nebula
● Galaxy
⦸ Just a Few Stars

Nunki
α φ
λ
M28

M19
Antares
α
M4

υ
τ

Mags
● 0th
● 1st
● 2nd
· 3rd
· 4th

ω
τ
M54
SAGITTARIUS
δ
γ

M62
−30°

M55
M70
M69
Kaus Australis
ε
η

M6
M7
Shaula
λ υ
κ
ε
μ
−40°

21h
20h
19h
18h
17h
16h
15h

MICROSCOPIUM
CORONA AUSTRALIS
SCORPIUS
LUPUS
CENTAURUS
Menkent
θ

γ δ

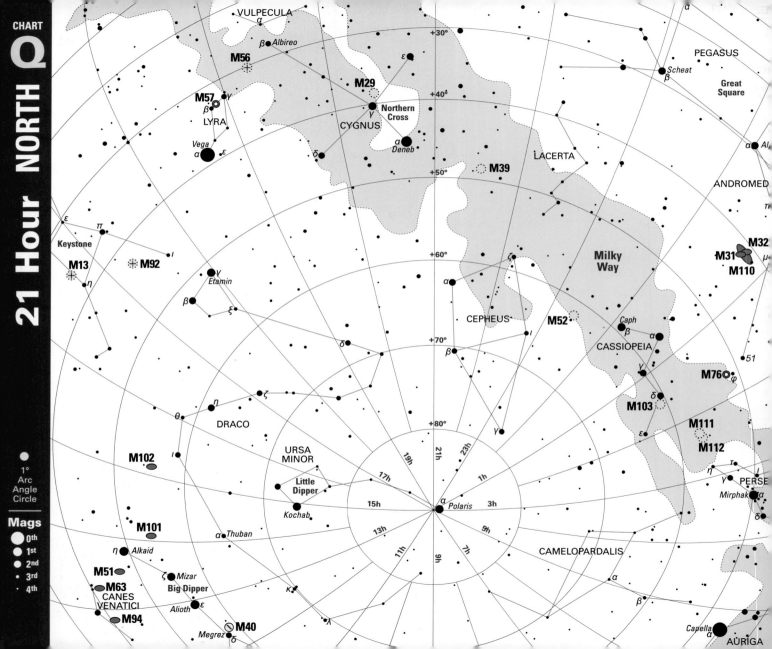

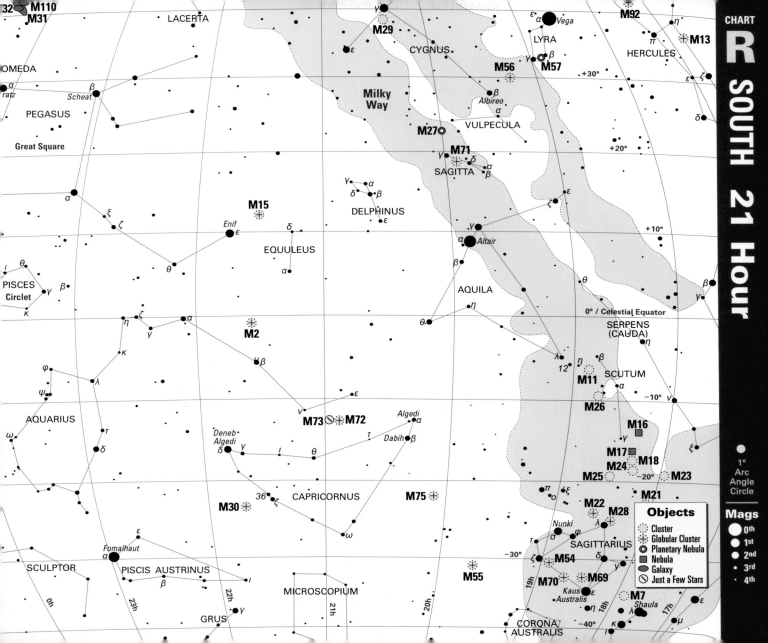

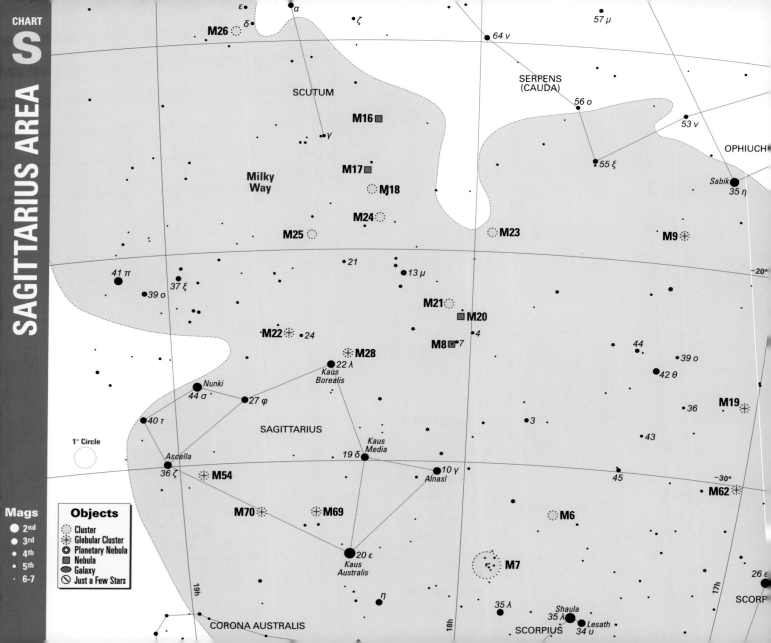

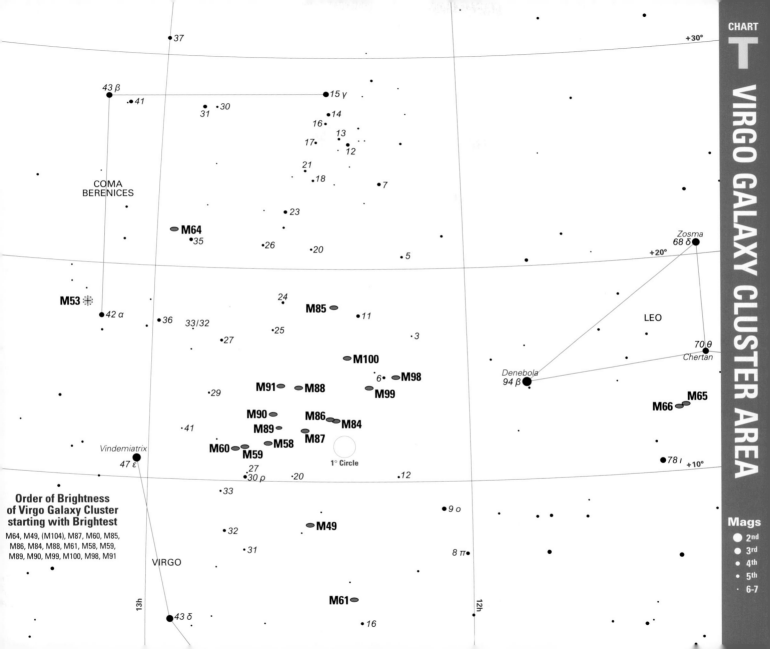

+30°

37

43 β

41

30

31

15 γ

14

16

13

17

12

COMA
BERENICES

21

18

7

23

M64

35

26

20

5

Zosma
68 δ

+20°

LEO

M53 ✳

42 α

36

33/32

24

M85

11

25

3

70 θ

Chertan

27

M100

29

M91

M88

6

M98

M99

Denebola
94 β

M90

M86

M84

M65

M89

M87

M66

M60

M58

78 ι

+10°

M59

Vindemiatrix

47 ε

27

30 ρ

20

12

1° Circle

33

**Order of Brightness
of Virgo Galaxy Cluster
starting with Brightest**

M64, M49, (M104), M87, M60, M85,
M86, M84, M88, M61, M58, M59,
M89, M90, M99, M100, M98, M91

32

M49

9 o

31

8 π

VIRGO

13h

43 δ

M61

16

12h

Mags

● 2nd

● 3rd

● 4th

• 5th

· 6-7

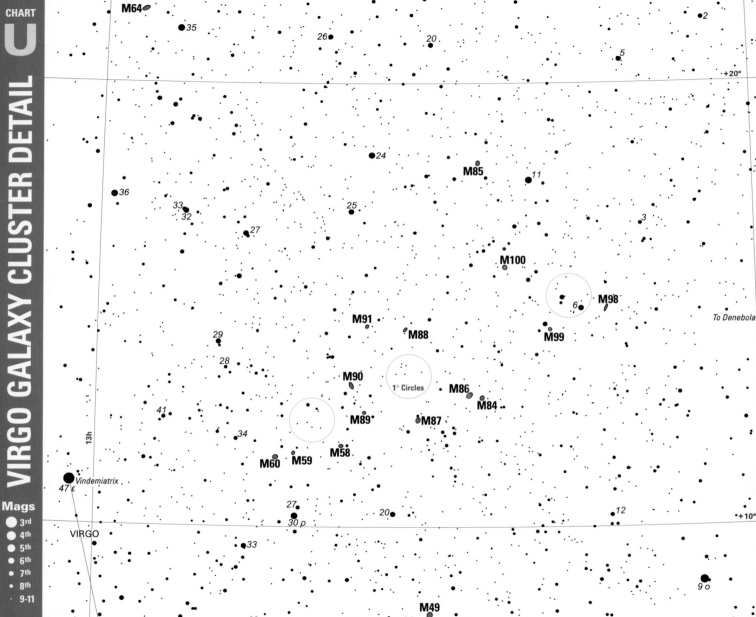

VIRGO GALAXY CLUSTER DETAIL

M64

35

26

20

2

5

+20°

24

M85

11

36

33

32

25

3

27

M100

6

M98

To Denebola

M91

M88

M99

29

28

M90

1° Circles

M86

M84

41

34

M89

M87

M58

M60 M59

M49

Vindemiatrix

47 ε

27

20

12

+10°

30 ρ

VIRGO

33

9 o

32

Mags
- 3rd
- 4th
- 5th
- 6th
- 7th
- 8th
- 9-11

13h

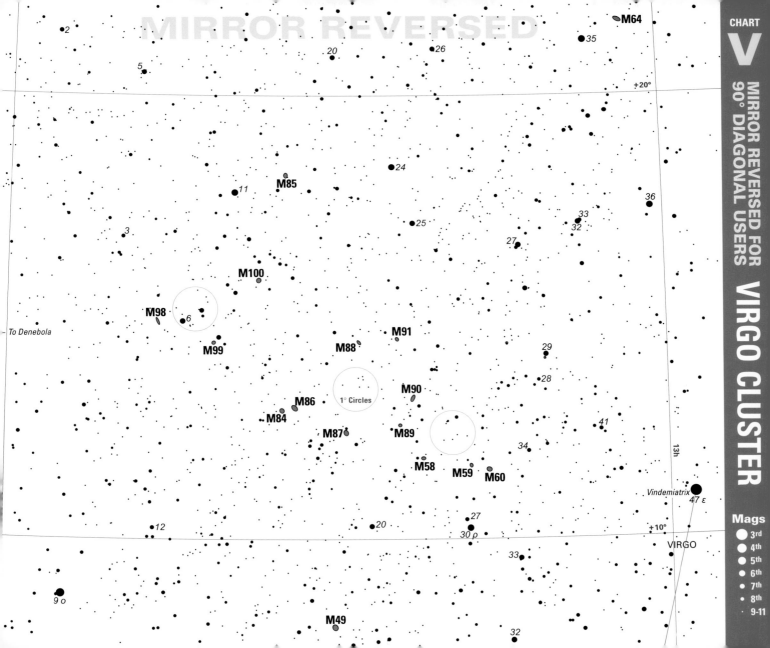

MIRROR REVERSED

CHART

V

MIRROR REVERSED FOR 90° DIAGONAL USERS

VIRGO CLUSTER

M64

2

35

20

26

5

+20°

24

M85

11

36

25

33

32

3

27

M100

M98

6

29

To Denebola

M99

M91

M88

28

M86

M90

1° Circles

M84

41

M87

M89

34

M58

M59

M60

12

27

30 ρ

13h

Vindemiatrix

47 ε

20

+10°

VIRGO

33

9 o

32

M49

Mags
● 3rd
● 4th
● 5th
● 6th
• 7th
• 8th
· 9-11

CATALOGUE
OF THE
NEBULAE AND STAR CLUSTERS

Observed in Paris, by Mr. Messier, at the Navy Observatory, Hotel Clugni, on Rue des Mathurins.

M̲r. Messier has most carefully observed the Nebulae & Star Clusters that can be seen over Paris' horizon; he determined their right ascension and their declination, and provided their diameters, along with detailed information on each of them: a publication that was lacking in the field of Astronomy.

He also goes into detail about the research he conducted on various nebulae whose discoveries are credited to several astronomers, but that he unsuccessfully found.

Mr. Messier's Catalogue of Nebulae & Star Clusters appeared in the volume of the Academy of Sciences, *year 1771, page 435.* He included at the end of his Report a very carefully traced drawing of the beautiful Orion Sword Nebula, with the stars it contains. This drawing will make it possible to check if, with the passing of time, it undergoes any changes. If you compared this drawing now with those of Messrs. Huygens, Picard, de Mairan & le Gentil, you would be surprised to find that it had been changed in such a way that it would be hard to believe it was the same nebula, based only on its shape. These drawings, rendered by Mr. le Gentil, can be seen in the 1759 volume of the Academy, *page 470, figure XXI.*

In regards to Mr. Messier's Catalogue presented here, we have added a large number of nebulae & star clusters he has discovered since his Report was printed & he has been reporting to us.

For each nebula, Mr. Messier assigned a number that can also be found on the following pages, along with details on each of the nebulae he observed.

The
Messier Objects

DATE des OBSERVATIONS.	Numéros des Nébuleuses.	ASCENSION DROITE.		DÉCLINAISON.	Diamètre en degrés & min.
		En Temps.	En Degrés.		
		H. M. S.	D. M. S.	D. M. S.	D. M
1758. Sept. 12	1.	5. 20. 2	80. 0. 33	21. 45. 27 B	
1760. Sept. 11	2.	21. 21. 8	320. 17. 0	1. 47. 0 A	0. 4
1764. Mai. 3	3.	13. 31. 25	202. 51. 19	29. 32. 57 B	0. 3
8	4.	16. 9. 8	242. 16. 56	25. 55. 40 A	0. 2½
23	5.	15. 6. 36	226. 39. 4	2. 57. 16 B	0. 3

N.ᵒˢ des Nébul.	Détails des Nébuleuses & des amas d'Étoiles. Les positions sont rapportées ci-contre.
1.	Nébuleuse au dessus de la corne méridionale du Taureau, ne contient aucune étoile; c'est une lumière blancheâtre, alongée en forme de la lumière d'une bougie, découverte en observant la Comète de 1758. Voyez la Carte de cette Comète, *Mém. Acad. année 1759, page 188*; observée par le Docteur Bévis vers 1731. Elle est rapportée sur l'*Atlas céleste* anglois.
2.	Nébuleuse sans étoile dans la tête du Verseau, le centre en est brillant, & la lumière qui l'environne est ronde; elle ressemble à la belle Nébuleuse qui se trouve entre la tête & l'arc du Sagittaire, elle se voit très-bien avec une lunette de deux pieds, placée sur le parallèle de α du Verseau. M. Messier a rapporté cette nébuleuse sur la Carte de la route de la Comète observée en 1759. *Mém. Acad. année 1760, page 464*. M. Maraldi avoit vu cette nébuleuse en 1746, en observant la Comète qui parut cette année.
3.	Nébuleuse découverte entre le Bouvier & un des Chiens de Chasse d'Hévélius, elle ne contient aucune étoile, le centre en est brillant & sa lumière se perd insensiblement, elle est ronde; par un beau ciel on peut la voir avec une lunette d'un pied; elle sera rapportée sur la Carte de la Comète observée en 1779. *Mémoires de l'Académie de la même année*. Revue le 29 Mars 1781, toujours très-belle.
4.	Amas d'étoiles très-petites; avec une foible lunette on le voit sous la forme d'une nébuleuse; cet amas d'étoiles est placé près d'*Antarès* & sur son parallèle. Observé par M. de la Caille, & rapporté dans son Catalogue. Revu le 30 Janvier & le 22 Mars 1781.
5.	Belle Nébuleuse découverte entre la Balance & le Serpent, près de l'étoile du Serpent, de sixième grandeur, la cinquième suivant le Catalogue de Flamsteed: elle ne contient aucune étoile; elle est ronde, & on la voit

U

Legend for Messier Objects

Original Messier Descriptions

These descriptions were taken from the French *Connaissance des Temps* (translated as *Knowledge of the Times*) 1784 edition. This was the third and last time Messier published his catalogue, with more entries than ever, totaling 103.

Messier's descriptions were translated by Isabelle Houthakker whom I have been fortunate to work with on several projects. She is an exceptional French translator because of her thoroughness and attention to detail. For these descriptions, I specifically asked her to keep the "flavor" of the printed text in *Connaissance des Temps*. This includes keeping any abbreviations, italics, symbols (like ampersands, that is, the "&") and contractions so that the translated English mimics the original French as closely as possible.

My notes provide additional information about names and facts in the descriptions.

NGC Summaries

The NGC Descriptions were taken from *NGC 2000.0*, edited by Roger W. Sinnott and published by Sky Publishing Corporation. The NGC descriptions that I present are an "expansion" of the actual descriptions which consist of a series of abbreviations and modifiers. For example, the actual NGC description for M1 reads: vB, vL, E 135°±, vglbM, r; = M1. You can decipher this by reading my expansion on page 108. Dreyer used approximately 125 abbreviations and modifiers. Keep in mind that Dreyer's descriptions date to observations prior to 1900.

Locations

The constellations and their three-letter abbreviation are noted along with coordinates calculated for the year 2000. Remember, none of these objects are actually "in" these constellations, but rather lie in their direction.

Observation Periods

These are the months when these objects are visible in the sky at around 8 p.m. and 4 a.m. Standard Time for most people in North America. Add one hour for Daylight Saving Time from April to October.

Facts

Facts **vary widely** from source to source. My sources were: *Sky Atlas 2000.0 Companion*, *NGC 2000.0*, *Sky & Telescope's Messier Card*, *The Messier Objects* by Stephen James O'Meara, *Observer's Handbook 2004* and articles in *Sky & Telescope* magazine. The physical dimensions of objects were calculated by using simple trigonometry based on their distances. Also see the definitions of Galaxy and Globular Cluster in the Glossary for supporting information on these objects.

Description of Messier Objects

These are my descriptions based on using a 4-inch diameter refractor in fairly dark locations. I kept them short in order to leave room for observers to form their own impressions. Remember, objects can look quite different depending on factors like atmospheric conditions, light pollution and altitude in the sky.

Locating Index

I chose to use the short and straightforward scale of **Easy**, **Fairly Easy**, **Slightly Challenging**, **Challenging and Difficult**. Any scale is relative, but I think this one will be intuitive after you have found several objects.

Identifying/Observing Index

Uses the same scale as the Locating Index.

Photographs

Each photograph is a 20-minute exposure on 400 speed negative film through a 4-inch refractor at a location with reasonably dark skies. The photographs are all at the same scale to provide convenient comparisons.

The photographs show the objects as if viewed with approximately 15x to 20x magnification held at normal reading distance.

There is an *arc degree* scale (see Glossary) along the left side of the photographs for measurements and a directional indicator pointing to celestial north.

Typical Eyepiece Fields of View (FOV) Scaled to the Photographs of the Messier Objects

This circle measures **1⅜ inches or 35mm** in diameter.

½° FOV is the same area that the Moon and Sun cover in the sky and can be obtained with an 8-inch SCT using a 20mm Plossl eyepiece.

Using the same scale as the photos, here is the largest that **Jupiter** will appear in the sky — measuring 50 arc seconds in diameter (50").

1° FOV can be obtained with an 8-inch SCT using a 40mm Plossl eyepiece. Any Plossl that provides a telescope magnification of about 50x will also provide a 1° FOV.

This circle measures **2¾ inches or 70mm** in diameter.

M1 Crab Nebula ✧ Supernova Remnant

Original Messier Description

from the 1784 edition of *Connaissance des Temps*

Observed 1758. Sept. 12

Nebula above the southern horn of Taurus, contains no stars; it's a whitish light, elongated in the shape of a candle's flame, discovered while observing the Comet of 1758. See the Map of this Comet. *Mém. Acad., year 1759, page 188;* observed by Dr. Bévis in about 1731. It's reported on the English *Celestial Atlas*.

Author's Notes: Messier had correspondence with Dr. John Bévis (1695–1771) from England, a physician and amateur astronomer whose star atlas *Uranographia Britannica* was published posthumously. This atlas indicated M1, M11, M13, M22, M31 and M35.
 Mém. Acad. refers to *Mémoires de l'Académie des Sciences* (Reports of the Academy of Sciences).

NGC Summary by J.L.E. Dreyer, circa 1888

NGC Number: 1952
Mag: 8.4 **Size:** 6'
NGC Description: Very bright, very large, extended along position angle of about 135°, very gradually becomes a little brighter in the middle, mottled.

Location

Constellation: Taurus (Tau)
Year 2000 Coordinates
RA: 5h 34.5m
Dec: +22° 01'

Observation Periods

Evenings 8 p.m. : November to April
Mornings 4 a.m. : August to January

Facts

Name: Crab Nebula
Type of Object: Supernova remnant
Magnitude: 8
Distance: 4,000 ly
Physical Size: 7 ly x 4.6 ly
Arc Degree Size in Sky: 6' x 4'
Other: The explosion which created this remnant occurred on July 4, 1054 and was recorded by the Chinese, Arabs and Japanese. It was visible for about 22 months, and for a while, could be seen during the day. The Crab Nebula is expanding at the rate of about 990 miles/ second. The last supernova explosion in our galaxy occurred in 1604.

Description of M1 using a 4-inch refractor at 48x

Shaped like a candle flame. With averted vision, it shows a brighter center that fades outward on opposite sides. Similar in size to M78 but brighter and more extended.

Locating Index: Easy because it is about 1° from the southern 2½ magnitude "horn" star, ζ *Tauri*, "toward" the star *Alnath*.

Identifying/Observing Index: Easy or **difficult** depending on the darkness of your skies because it gets washed out easily by light pollution. I grew up in Milwaukee and was unable to see this gem with a 10-inch telescope from our northside home.

TO LOCATE Refer to Star Charts **B, F** & **H**

The Crab Nebula caught Messier's attention because its shape resembles some comets. It was this nebula that inspired him to start cataloguing deep sky objects.

For easy comparison, all photographs are shown at the same scale, measuring 2.3° x 1.3°.

M2 ✧ Globular Cluster

Original Messier Description

from the 1784 edition of *Connaissance des Temps*

Observed 1760. Sept. 11

Nebula without stars in the head of Aquarius, its center is brilliant & the light surrounding it is circular. It resembles the beautiful nebula located between the head & bow of Sagittarius. You can see it very well with a two-foot telescope placed on the parallel of the α of Aquarius. Mr. Messier reported this nebula on the Chart of the track of the Comet observed in 1759. *Mém. Acad. year 1760, page 464.* Mr. Maraldi had seen this nebula in 1746, while observing the Comet which appeared that year.*

NGC Summary by J.L.E. Dreyer, circa 1888

NGC Number: 7089
Mag: 6.5 **Size:** 12.9'
NGC Description: Very remarkable!! Globular cluster of stars, bright, very large, gradually becomes pretty much brighter in the middle, well resolved, contains extremely faint stars.

Location

Constellation: Aquarius (Aqr)

Year 2000 Coordinates
RA: 21h 33.5m
Dec: −0° 49'

Observation Periods

Evenings 8 p.m. : August to December
Mornings 4 a.m. : April to August

Facts

Name: No common name
Type of Object: Globular cluster
Magnitude: 6½
Distance: 40,000 ly
Physical Size: 151 ly in diameter
Arc Degree Size in Sky: 13' in diameter
Other: The stars in this globular cluster are very tightly compacted and are rated II on a I to XII scale where I is the most compact. See *Globular Cluster* in the glossary for more information.

---

*Author's Note: Jean-Dominique Maraldi (1709–1788) was originally from Italy, but later moved to Paris. He was related to Cassini. In Paris, he worked on measuring longitude using Jupiter's moons, observed comets and assisted in the publication of 25 annual issues of *Connaissance des Temps*.

Description of M2 using a 4-inch refractor at 48x

Overall, pretty bright and large. Very noticeable. Similar to M15. Bright center that fades outward. In an area sparse with stars.

Locating Index: **Fairly easy** since it forms the right angle of a triangle with the two brightest stars in the handle of Aquarius, α and β *Aquarii*.

Identifying/Observing Index: **Easy** since this jewel is bright and thus stands out when you come across it.

TO LOCATE
Refer to
Star Charts
A,
D
& **R**

ARC SCALE
1° —
3/4° —
1/2° —
25' —
20' —
15' —
10' —
5' —
0' —

N

M2 is a surprising joy in an area devoid of brighter stars.

For easy comparison, all photographs are shown at the same scale, measuring 2.3° x 1.3°.

M3 ✧ Globular Cluster

Original Messier Description

from the 1784 edition of *Connaissance des Temps*

Observed 1764. May 3

Nebula discovered between Boötes & one of Hevelius' Canes Venatici; it contains no stars; its center is brilliant & its light gradually fades away; it's round. In a good sky, it can be seen through a one-foot telescope. It's reported on the Chart of the Comet observed in 1779. *Mémoires de l'Académie of the same year.* Observed again on March 29, 1781, still very beautiful.

Author's Note: Johan Hevelius (1611–1687) was a wealthy German brewer and politician who was also an astronomer. Aside from building several observatories, he published his observations of comets, maps of the Moon, and posthumously, a star catalogue and atlas.

NGC Summary   by J.L.E. Dreyer, circa 1888

NGC Number: 5272
Mag: 6.4   **Size:** 16.2'
NGC Description: Very remarkable!! Globular cluster of stars, extremely bright, very large, very suddenly becomes much brighter in the middle, contains stars of 11th magnitude and fainter.

Location

Constellation: Canes Venatici (CVn)
Year 2000 Coordinates
RA: 13h 42.2m
Dec: +28° 23'

Observation Periods

Evenings 8 p.m. : March to September
Mornings 4 a.m. : November to May

Facts

Name: No common name
Type of Object: Globular cluster
Magnitude: 6.2
Distance: 35,000 ly
Physical Size: 165 ly in diameter
Arc Degree Size in Sky: 16.2' in diameter
Other: Contains about 50,000 stars. The compactness of this globular cluster is considered average. Its age is estimated at 6,500,000,000 years.

Description of M3 using a 4-inch refractor at 48x

Magnificent globular cluster with a very bright extended core. Visually, just as impactful as M13. Can see individual stars. This is one you don't want to pass up. Try finding it with binoculars.

Locating Index: Slightly challenging because it is not located near any obvious stars. To find it, start by pointing your telescope halfway between *Arcturus* and the bright, "corner" star of Canes Venatici, *α Canum Venaticorum*.

Identifying/Observing Index: Easy because this globular really pops out when you come across it.

M3

TO LOCATE
Refer to
Star Charts
**A,
L**
& **N**

ARC SCALE

1° —
3/4° —
1/2° —
25' —
20' —
15' —
10' —
5' —
0' —

Visually, M3 is one of the bigger and brighter globulars in the sky.

For easy comparison, all photographs are shown at the same scale, measuring 2.3° x 1.3°.

M4 Cat's Eye ✦ Globular Cluster

Original Messier Description

from the 1784 edition of *Connaissance des Temps*

Observed 1764. May 8

Cluster of very small stars: with a weak telescope, it looks like a nebula; this cluster of stars is situated near *Antares* & on its parallel. Observed by Mr. de la Caille, & reported in his Catalogue. Observed again on January 30 & March 22, 1781.

Author's Notes: Messier's phrase "very small stars" means very faint stars. Many of the original NGC descriptions also use the words small and large in reference to faint and bright stars.
  French Abbot Nicholas Louis de La Caille (1713–1762) complied a list of 42 deep sky objects during his 1751–1752 journey to the Cape of Good Hope.

Location

Constellation: Scorpius (Sco)

Year 2000 Coordinates
RA: 16h 23.6m
Dec: −26° 32'

Observation Periods

Evenings 8 p.m. : June to September
Mornings 4 a.m. : February to May

Facts

Name: Cat's Eye
Type of Object: Globular cluster
Magnitude: 5.9
Distance: 14,000 ly
Physical Size: 107 ly in diameter
Arc Degree Size in Sky: 26.3' in diameter
Other: The compactness of this globular cluster is much looser than average. Its age is estimated at 10,000,000,000 years.

NGC Summary by J.L.E. Dreyer, circa 1888

NGC Number: 6121
Mag: 5.9   **Size:** 26.3'
NGC Description: Cluster, 8 or 10 bright stars in a line, with 5 stars that are well resolved.

Description of M4 using a 4-inch refractor at 48x

A most visually unique globular cluster because it has a line of bisecting stars that is easy to see. Loosely concentrated center, dominated by the "line." Can see individual stars, and by using averted vision, you can see many more.

Locating Index: Easy because it is about one degree "west" of the star *Antares*.

Identifying/Observing Index: Fairly easy. I say "fairly" because even though this globular is large, it is not as bright as it appears in the picture. In poor skies, I have passed it by because it can fade into light-polluted backgrounds. Probably a wonderful sight from the southern hemisphere.

TO LOCATE Refer to Star Charts **A, N** & **P**

The bright star *Antares* is at the left and the fainter 3rd magnitude σ *Scorpii* is at the right. A smaller and fainter globular cluster, NGC 6144, is noted at the top of the picture.

For easy comparison, all photographs are shown at the same scale, measuring 2.3° x 1.3°.

M5 ✦ Globular Cluster

Original Messier Description

from the 1784 edition of *Connaissance des Temps*

Observed 1764. May 23

Beautiful nebula discovered between Libra & Serpens, near the star of Serpens, sixth magnitude, fifth according to the Flamsteed Catalogue. It contains no stars; it's round & you can see it very well, in a good sky, with an ordinary one-foot telescope. Mr. Messier reported it on the Chart of the Comet of 1763. *Mém. Acad. year 1774, page 40.* Observed again on Sept. 5, 1780, January 30 & March 22, 1781.

Author's Note: Additional information about John Flamsteed and his numbering system can be found on page 80 and in the Glossary.

NGC Summary by J.L.E. Dreyer, circa 1888

NGC Number: 5904
Mag: 5.8 **Size:** 17.4'
NGC Description: Very remarkable!! Globular cluster of stars, very bright, large, extremely compressed in the middle, contains stars of magnitude 11 to 15.

Location

Constellation: Serpens Caput* (Ser)
Year 2000 Coordinates
RA: 15h 18.6m
Dec: +2° 05'

Observation Periods

Evenings 8 p.m. : May to September
Mornings 4 a.m. : January to May

Facts

Name: No common name
Type of Object: Globular cluster
Magnitude: 5.7
Distance: 26,000 ly
Physical Size: 132 ly in diameter
Arc Degree Size in Sky: 17.4' in diameter
Other: Slightly oval in shape in northeast/southwest direction. Compactness is a little tighter than average. Its age is estimated at 13,000,000,000 years.

---

* Serpens is the only constellation with discontinuous boundaries because it straddles Ophiuchus (known for holding a snake that extends across him). Serpens west of Ophiuchus is often referred to as Serpens Caput (head of the snake) and the portion to the east as Serpens Cauda (tail of the snake).

Description of M5 using a 4-inch refractor at 48x

A show-stopper and in league with the best. It's big, bright and very extended. Can see many individual stars with averted vision. There is one fairly bright star near its center. Within ½ degree of the 5th magnitude star *5 Serpentis*.

Locating Index: Slightly challenging because it is not located near any obvious stars. To find it, you can use several triangle shapes formed with the stars of Serpens and/or Virgo.

Identifying/Observing Index: Easy to spot when you come across it because it is very large and bright.

ARC SCALE

1° —
3/4° —
1/2° —
25' —
20' —
15' —
10' —
5' —
0' —

5

TO LOCATE
Refer to Star Charts
A,
N
& **P**

This show-stopper is big, bright and near a 5th magnitude star, *5 Serpentis*, noted.

For easy comparison, all photographs are shown at the same scale, measuring 2.3° x 1.3°.

M6 Butterfly Cluster ✧ Open Cluster

Original Messier Description

from the 1784 edition of *Connaissance des Temps*

Observed 1764. May 23

Cluster of small stars between the bow of Sagittarius & the tail of Scorpius. To the naked eye it seems to be a nebulosity without stars, but using even the smallest instrument to examine it, you'll see a cluster of small stars.

Location

Constellation: Scorpius (Sco)

Year 2000 Coordinates
RA: 17h 40.1m
Dec: −32° 13'

Observation Periods

Evenings 8 p.m. : July to September
Mornings 4 a.m. : March to May

Facts

Name: Butterfly Cluster
Type of Object: Open cluster
Magnitude: 4.2
Distance: 1,500 ly
Physical Size: Spans 6.6 ly across its longest length
Arc Degree Size in Sky: Extends to 15'
Other: This cluster contains about 330 stars ranging in magnitude from 6.2 to 14 and fainter. About 60 of these stars are magnitude 11 and brighter. Its age is estimated at 51,000,000 years.

NGC Summary by J.L.E. Dreyer, circa 1888

NGC Number: 6405
Mag: 4.2   **Size:** 15'
NGC Description: Cluster, large, irregularly round, little compressed, contains stars of 7th magnitude, 10th and fainter.

Description of M6 using a 4-inch refractor at 48x

Big and bright compared to most clusters but not as spectacular as M7 south of it. Many of its stars appear to be of similar magnitude and blue-white in color; however, its brightest star has a red-orange coloring.

Locating Index: Easy in dark skies because it can be seen with the naked eye. **Slightly challenging** to find in light-polluted skies because there are no conspicuous bright reference stars. However, it forms the apex of a shallow triangle with *Shaula* and *γ Sagittarii*.

Identifying/Observing Index: Easy because it "pops" when you come across it.

TO LOCATE
Refer to
Star Charts
A,
P
& **S**

The bright, red-orange star in this cluster is indicated by the letter R. Also indicated in the picture is Tr 28 (Tr stands for Trumpler), a smaller and fainter cluster not quite a degree away from M6.

For easy comparison, all photographs are shown at the same scale, measuring 2.3° x 1.3°.

M7 ✦ Open Cluster

Original Messier Description

from the 1784 edition of *Connaissance des Temps*

Observed 1764. May 23

Cluster of stars considerably larger than the preceding one. This cluster looks like a nebulosity to the naked eye; it's not far from the preceding one, located between the bow of Sagittarius & the tail of Scorpius.

Location

Constellation: Scorpius (Sco)

Year 2000 Coordinates
RA: 17h 53.9m
Dec: –34° 49'

Observation Periods

Evenings 8 p.m. : July to September
Mornings 4 a.m. : March to May

Facts

Name: No common name
Type of Object: Open cluster
Magnitude: 3.3
Distance: 800 ly
Physical Size: Spans 20 ly across its longest length
Arc Degree Size in Sky: Extends to 80'
Other: Totals about 80 stars with the brightest star shining at magnitude 5.6. Its age is estimated at 220,000,000 years.

NGC Summary by J.L.E. Dreyer, circa 1888

NGC Number: 6475
Mag: 3.3 **Size:** 80'
NGC Description: Cluster, very bright, pretty round but with irregularities, little compressed, contains stars of magnitude 7 to 12.

Description of M7 using a 4-inch refractor at 48x

A beautiful and very large cluster that fills the whole eyepiece field of view. Rich with many bright members scattered throughout. At its center, where there is the highest concentration of stars, I almost see the shape of the constellation Hercules.

Locating Index: **Easy** in dark skies because it can be seen with the naked eye. **Slightly challenging** to find in light-polluted skies because there are no conspicuous bright reference stars. However, it forms the apex of a shallow triangle with *Shaula* and γ *Sagittarii*.

Identifying/Observing Index: **Easy** because it "pops" when you come across it.

M7

TO LOCATE
Refer to
Star Charts
A,
P
& S

1°—
ARC SCALE
3/4°—
1/2°—
25'—
20'—
15'—
10'—
5'—
0'—

M7 is in a thick part of the Milky Way. In dark skies, it can very easily be seen with the naked eye. For the most part, all of the "brighter" stars in this picture are a part of this cluster.

For easy comparison, all photographs are shown at the same scale, measuring 2.3° x 1.3°.

M8 Lagoon Nebula ✧ Nebula

Original Messier Description

from the 1784 edition of *Connaissance des Temps*

Observed 1764. May 23

Cluster of stars that looks like a nebula when observed through an ordinary three-foot telescope; however, with an excellent instrument you will only see a large number of small stars there; near this cluster there's a fairly bright star surrounded by very faint light; this is the ninth star of Sagittarius, seventh magnitude according to Flamsteed: this cluster appears as an elongated shape extending from the northeast to the southwest, between the bow of Sagittarius & the right foot of *Ophiuchus*.

NGC Summary by J.L.E. Dreyer, circa 1888

NGC Number: 6523
Mag: 5.8 **Size:** 90'
NGC Description: A magnificent object!!! Very bright, extremely large, extremely irregular and faint, with a large cluster.

Location

Constellation: Sagittarius (Sgr)

Year 2000 Coordinates
RA: 18h 03.8m
Dec: −24° 23'

Observation Periods

Evenings 8 p.m. : July to October
Mornings 4 a.m. : March to June

Facts

Name: Lagoon Nebula
Type of Object: Emission nebula of the kind from which new stars are born
Magnitude: 6
Distance: 4,800 ly
Physical Size: 126 ly x 56 ly
Arc Degree Size in Sky: 90' x 40'
Other: This nebula produces its light through fluorescence: its atoms are stimulated to give off light by the highly energetic light of the 6th magnitude star *9 Sagittarii*. This process of producing light is similar to that of neon signs.

Description of M8 using a 4-inch refractor at 48x

Large but overall the nebulosity is fainter than you might expect. M17 farther north is the brightest nebula in this area. M8 appears as two parts. The large patch is lit by the bright star *9 Sagittarii* just off center. To the side of the large patch, there is a star cluster embedded in nebulosity.

Locating Index: Fairly easy in dark skies because to the naked eye, it appears as a bright patch of the Milky Way. Otherwise, it is found by pointing your telescope at the end of a one length western extension of the line between φ and λ *Sagittarii*.

Identifying/Observing Index: Easy because the cluster and associated nebulosity is quite large.

TO LOCATE
Refer to
Star Charts
**A,
P
& S**

ARC SCALE

1°

3/4°

1/2°

25'

20'

15'

10'

5'

0'

N

6553

The Lagoon Nebula is very large, and details within its nebulosity can be seen with small telescopes. *9 Sagittarii* is located in the brightest part of the nebula. Globular cluster NGC 6553 is noted.

For easy comparison, all photographs are shown at the same scale, measuring 2.3° x 1.3°.

M9 ✧ Globular Cluster

Original Messier Description

from the 1784 edition of *Connaissance des Temps*

Observed 1764. May 28

Nebula without stars in the right leg of *Ophiuchus*; it's round & its light is faint. Observed again on March 22, 1781.

Location

Constellation: Ophiuchus (Oph)

Year 2000 Coordinates
RA: 17h 19.2m
Dec: −18° 31'

Observation Periods

Evenings 8 p.m. : June to October
Mornings 4 a.m. : February to May

Facts

Name: No common name
Type of Object: Globular cluster
Magnitude: 7.7
Distance: 27,000 ly
Physical Size: 73 ly in diameter
Arc Degree Size in Sky: 9.3' in diameter
Other: Located on the border of a dark nebula known as Barnard 64.

NGC Summary by J.L.E. Dreyer, circa 1888

NGC Number: 6333
Mag: 7.9   **Size:** 9.3'
NGC Description: Globular cluster of stars, bright, large, round, extremely compressed in the middle, well resolved, contains stars of magnitude 14.

Description of M9 using a 4-inch refractor at 48x

Similar in size to M62 farther south but much fainter because, overall, it is much more diffused. Center is better glimpsed using averted vision. I could not discern the dark nebula next to it.

Locating Index: **Fairly easy** because it is not too far from the bright star *Sabik* and forms triangles with several nearby stars.

Identifying/Observing Index: **Fairly easy** in dark skies but will prove to be more **challenging** in light-polluted ones because its overall brightness is low.

TO LOCATE
Refer to
Star Charts
**A,
P**
& **S**

1° —

X

0025 75540

20' —
15' —
10' —
5' —
0' —

6356

As pictured, M9 lies to the left of a dark nebula known as Barnard 64. Dark nebulae are seen as silhouettes because they block starlight behind them. The fainter globular NGC 6356 is noted.

For easy comparison, all photographs are shown at the same scale, measuring 2.3° x 1.3°.

M 10 ✧ Globular Cluster

Original Messier Description

from the 1784 edition of *Connaissance des Temps*

Observed 1764. May 29

Nebula without stars in the belt of *Ophiuchus*, near this constellation's thirtieth star, sixth magnitude according to Flamsteed. This nebula is beautiful & round. It was hard to see with an ordinary three-foot telescope. Mr. Messier reported it on the second Chart of the track of the Comet of 1769. *Mém. Acad. year 1775, plate IX.* Observed again on March 6, 1781.

NGC Summary by J.L.E. Dreyer, circa 1888

NGC Number: 6254
Mag: 6.6   **Size:** 15.1'
NGC Description: Remarkable! Globular cluster of stars, bright, very large, round, gradually becomes very much brighter in the middle, well resolved, contains stars of magnitude 10 to 15.

Location

Constellation: Ophiuchus (Oph)

Year 2000 Coordinates
RA: 16h 57.1m
Dec: −4° 06'

Observation Periods

Evenings 8 p.m. : June to October
Mornings 4 a.m. : February to June

Facts

Name: No common name
Type of Object: Globular cluster
Magnitude: 6.6
Distance: 20,000 ly
Physical Size: 88 ly in diameter
Arc Degree Size in Sky: 15.1' in diameter
Other: This globular is just a little looser than average in compactness.

Description of M10 using a 4-inch refractor at 48x

A very pretty, large and bright globular that is similar in size to M5, slightly smaller and fainter than M13 but certainly in the same league as these two beauties! The core is bright and more extended than others.

Locating Index: Challenging because it is not near any bright stars. It can be found by aiming the telescope on a line ⅓ the distance from ζ *Ophiuchi* and *Rasalhague*. It also forms a right triangle with ζ *Ophiuchi* and *Sabik*.

Identifying/Observing Index: Easy. Pops out when you come across it because of its size and brightness.

TO LOCATE
Refer to
Star Charts
A,
N
& **P**

M10 is one of the brighter globulars in the sky.

For easy comparison, all photographs are shown at the same scale, measuring 2.3° x 1.3°.

M11 Wild Duck Cluster ✧ Open Cluster

Original Messier Description

from the 1784 edition of *Connaissance des Temps*

Observed 1764. May 30

Cluster of a great number of small stars, near the star *K of Antinoüs*, which can only be seen with good instruments. With an ordinary three-foot telescope, it looks like a Comet. This cluster is suffused with a faint light. In this cluster there is an eighth magnitude star. Mr. Kirch observed it in 1681. *Transact. Philos. No. 347, page 390.* It's reported in the English *Great Atlas*.

Author's Notes: Aquila has replaced the constellation Antinoüs. *K of Antinoüs* is the 5th magnitude star *η Scuti* which is indicated on charts P and R. Bayer used Roman letters to designate stars after he exhausted the Greek alphabet — only a few holdovers exist today.

Gottfried Kirch (1639–1710) of Germany discovered M11 and M5 in 1702. He learned astronomy from Hevelius.

NGC Summary by J.L.E. Dreyer, circa 1888

NGC Number: 6705
Mag: 5.8 **Size:** 14'
NGC Description: Remarkable! Cluster, very bright, large, irregularly round, rich in stars, contains a star of 9th magnitude, contains stars of 11th magnitude and fainter.

Location

Constellation: Scutum (Sct)
Year 2000 Coordinates
RA: 18h 51.1m
Dec: −6° 16'

Observation Periods

Evenings 8 p.m. : July to October
Mornings 4 a.m. : March to June

Facts

Name: Wild Duck Cluster
Type of Object: Open cluster
Magnitude: 5.8
Distance: 5,600 ly
Physical Size: 23 ly at its widest
Arc Degree Size in Sky: 14'
Other: Contains about 200 stars, the brightest shining at magnitude 8. Its age is estimated at 220,000,000 years.

Description of M11 using a 4-inch refractor at 48x

One of my favorites. A beautiful cluster that almost looks like a loose globular. It has one bright star with many fainter stars fanning around it in a lopsided, blotchy fashion (like a leading duck within a flock). Use higher magnifications to see its shape and stars better.

Locating Index: Challenging because it is not near any conspicuously bright stars. Use the 3rd and 4th magnitude stars, like λ, 12, η and β at the bottom of Aquila and the top of Scutum to triangulate its position.

Identifying/Observing Index: Easy when you come across it because it is fairly bright.

TO LOCATE
Refer to
Star Charts
A,
P
& **R**

1°—
ARC SCALE
3/4°—
1/2°—
25'—
20'—
15'—
10'—
5'—
0'—

N

At first glance with lower magnifications, M11 may look like a globular cluster, but the stars quickly resolve with higher magnification. This "sprinkle" of like-magnitude stars is absolutely beautiful.

For easy comparison, all photographs are shown at the same scale, measuring 2.3° x 1.3°.

M 12 ✦ Globular Cluster

Original Messier Description

from the 1784 edition of *Connaissance des Temps*

Observed 1764. May 30

Nebula discovered in Serpens, between the arm & left side of *Ophiuchus*. This nebula contains no stars; it's round & its light is faint; near this nebula is a ninth magnitude star. Mr. Messier reported it on the second Chart of the Comet observed in 1769. *Mém. Acad. 1775, pl. IX.* Observed again on March 6, 1781.

NGC Summary by J.L.E. Dreyer, circa 1888

NGC Number: 6218
Mag: 6.6   **Size:** 14.5'
NGC Description: Very remarkable!! Globular cluster of stars, very bright, very large, irregularly round, gradually becomes much brighter in the middle, well resolved, contains stars of 10th magnitude and fainter.

Location

Constellation: Ophiuchus (Oph)

Year 2000 Coordinates
RA: 16h 47.2m
Dec: −1° 57'

Observation Periods

Evenings 8 p.m. : June to October
Mornings 4 a.m. : February to June

Facts

Name: No common name
Type of Object: Globular cluster
Magnitude: 6.7
Distance: 24,000 ly
Physical Size: 101 ly in diameter
Arc Degree Size in Sky: 14½' in diameter
Other: The compactness of the stars in this globular cluster is looser than average.

Description of M12 using a 4-inch refractor at 48x

About the same size as M10 but fainter and not quite as symmetrical. On initial inspection, this almost appears like an open cluster. Several bright 10th magnitude stars "lie" on top of this cluster, one close to its center.

Locating Index: Slightly challenging because it is not near any conspicuously bright stars. It forms various triangles with the southern and western stars of Ophiuchus.

Identifying/Observing Index: Easy. It pops out when you come across it because of its size and brightness.

TO LOCATE
Refer to
Star Charts
A,
N
& **P**

ARC SCALE

1° —
3/4° —
1/2° —
25' —
20' —
15' —
10' —
5' —
0' —

Although M12 is big and fairly bright, its next-door neighbor M10 is even bigger and brighter.

For easy comparison, all photographs are shown at the same scale, measuring 2.3° x 1.3°.

M13 Great Hercules Cluster ✧ Globular Cluster

Original Messier Description

from the 1784 edition of *Connaissance des Temps*

Observed 1764. Jun. 1

Nebula without stars, discovered in the belt of Hercules. It's round & brilliant, the center brighter than the edges, you can see it with a one-foot telescope. It's near two stars, both 8th magnitude, one above & the other below: the nebula was determined by comparing it to ε of Hercules. Mr. Messier reported it on the chart of the Comet of 1779, inserted in the Mémoires de l'Académie, of the *year 1784.* Observed by Halley in 1714. Observed again on Jan. 5 & 30, 1781. It's reported in the English *Celestial Atlas*.

NGC Summary by J.L.E. Dreyer, circa 1888

NGC Number: 6205
Mag: 5.9 **Size:** 16.6'
NGC Description: Very remarkable!! Globular cluster of stars, extremely bright, very rich in stars, very gradually becomes extremely compressed in the middle, contains stars of 11th magnitude and fainter.

Location

Constellation: Hercules (Her)

Year 2000 Coordinates
RA: 16h 41.7m
Dec: +36° 28'

Observation Periods

Evenings 8 p.m. : April to November
Mornings 4 a.m. : December to July

Facts

Name: Great Hercules Cluster
Type of Object: Globular cluster
Magnitude: 5.8
Distance: 21,000 ly
Physical Size: 104 ly in diameter
Arc Degree Size in Sky: 17' in diameter
Other: Considered by many to be the best globular cluster north of the celestial equator. Although its discovery is attributed to Edmond Halley in 1714, this cluster is visible to the naked eye. Contains about 500,000 stars, compacted a little tighter than average. Its age is estimated at 14,000,000,000 years.

Description of M13 using a 4-inch refractor at 48x

Very big. Its very bright core extends all the way to its "edge," then fades off gradually. I can just see individual stars with direct vision but many more with averted vision.

Locating Index: Easy because it is between two corner stars that make up the Keystone in Hercules. It is about ⅓ of the way down from the most northern, right-side corner star.

Identifying/Observing Index: Easy when you come across it because it "pops." In very dark skies, try to see it with your naked eyes.

TO LOCATE
Refer to
Star Charts
A,
O
& **P**

M13 is one of the biggest and brightest globulars in the sky. The small and faint galaxy NGC 6207 is noted.

For easy comparison, all photographs are shown at the same scale, measuring 2.3° x 1.3°.

M 14 ✧ Globular Cluster

Original Messier Description

from the 1784 edition of *Connaissance des Temps*

Observed 1764. Jun. 1

Nebula without stars, discovered in the drapery that goes across the right arm of *Ophiuchus*, & located on the parallel of ζ of Serpens; this nebula is not large, its light is faint, however, you can see it through an ordinary three-&-a-half-foot telescope. It's round; near it there's a small ninth magnitude star; its position was determined by comparing it to γ of *Ophiuchus*, & Mr. Messier reported its position on the Chart of the Comet of 1769. *Mémoires de l'Académie, year 1775, plate IX.* Observed again on March 22, 1781.

NGC Summary by J.L.E. Dreyer, circa 1888

NGC Number: 6402
Mag: 7.6 **Size:** 11.7'
NGC Description: Remarkable! Globular cluster of stars, bright, very large, round, extremely rich in stars, very gradually becomes much brighter in the middle, well resolved, contains stars of magnitude 15.

Location

Constellation: Ophiuchus (Oph)
Year 2000 Coordinates
RA: 17h 37.6m
Dec: –3° 15'

Observation Periods

Evenings 8 p.m. : June to October
Mornings 4 a.m. : February to June

Facts

Name: No common name
Type of Object: Globular cluster
Magnitude: 7.6
Distance: 29,000 ly
Physical Size: 101 ly in diameter
Arc Degree Size in Sky: 12' in diameter
Other: The trio of M10, M12 and M14 are all in the same class, but M14 ranks as faintest and smallest. The compactness of this globular is looser than average, but it is more compact than M12 and less than M10.

Description of M14 using a 4-inch refractor at 48x

Big and bright, appearing very much like a "cotton ball," because overall its luminosity gradually decreases. Although I can't see individual stars, with averted vision there does appear to be a higher concentration of stars at its center.

Locating Index: Slightly challenging because it is not near any bright stars. However, it is close to two 4½ magnitude stars which form a right-angle triangle with this globular.

Identifying/Observing Index: Easy because it is big and bright, and thus will "pop" when you come across it.

ARC SCALE

1° —
3/4° —
1/2° —
25' —
20' —
15' —
10' —
5' —
0' —

TO LOCATE Refer to Star Charts **A, N** & **P**

M14 is in the small class as M10 and M12 even though it is fainter and smaller.

For easy comparison, all photographs are shown at the same scale, measuring 2.3° x 1.3°.

M15 Great Pegasus Cluster ✦ Globular Cluster

Original Messier Description

from the 1784 edition of *Connaissance des Temps*

Observed 1764. Jun. 3

Nebula without stars, between the heads of Pegasus & Equuleus; it's round, its center is brilliant, its position determined by comparing it to δ of Equuleus. Mr. Maraldi talks about this nebula in the *Mémoires de l'Académie of 1746:* "I saw," said he, "between the star ε of Pegasus & β of Equuleus, a fairly bright nebulous star that's composed of many stars; its right ascension is 319^d 27' 6", & its northern declination is +11^d 2' 22"."

Author's Note: The RA reported above uses arc degrees instead of hours. I have seen this alternative measurement system side-by-side with the hour system on setting circles made as late as the 1970s.

NGC Summary by J.L.E. Dreyer, circa 1888

NGC Number: 7078
Mag: 6.4 **Size:** 12.3'
NGC Description: Remarkable! Globular cluster of stars, very bright, very large, irregularly round, very suddenly becomes much brighter in the middle, well resolved, contains stars that are very faint.

Location

Constellation: Pegasus (Peg)
Year 2000 Coordinates
RA: 21h 30.0m
Dec: +12° 10'

Observation Periods

Evenings 8 p.m. : July to January
Mornings 4 a.m. : March to September

Facts

Name: Great Pegasus Cluster
Type of Object: Globular cluster
Magnitude: 6.2
Distance: 34,000 ly
Physical Size: 122 ly in diameter
Arc Degree Size in Sky: 12.3' in diameter
Other: On "top" and at the very center of this globular is a 15th magnitude planetary nebula. Visually, it cannot be seen with small telescopes, but a hint of it appeared on my negative. Tighter than average compactness.

Description of M15 using a 4-inch refractor at 48x

A very nice example of a globular cluster that dominates the autumn skies. It has an intense, starlike center that is comparable in brightness to a star about 15 arc minutes away.

Locating Index: Easy because it is not far from the star *Enif* in Pegasus and almost on the same line extending outward from θ *Pegasi* and *Enif*.

Identifying/Observing Index: Easy because it is big and bright and "pops" when you come across it.

TO LOCATE
Refer to
Star Charts
A,
D
& **R**

1°—

ARC SCALE

3/4°—

1/2°—
25'—
20'—
15'—
10'—
5'—
0'—

N

M15 is a nice surprise that is easy to find.

For easy comparison, all photographs are shown at the same scale, measuring 2.3° x 1.3°.

M 16 Eagle Nebula ✧ Nebula & Open Cluster

Original Messier Description

from the 1784 edition of *Connaissance des Temps*

Observed 1764. Jun. 3

Cluster of small stars, suffused with a faint glow, near the tail of Serpens, not far from the parallel of that constellation's ζ; with a weak telescope, this cluster looks like a nebula.

Location

Constellation: Serpens Cauda (Ser)

__Year 2000 Coordinates__
RA: 18h 18.8m
Dec: –13° 47'

Observation Periods

Evenings 8 p.m. : July to October
Mornings 4 a.m. : March to June

Facts

Name: Eagle Nebula, the Ghost, or Star Queen Nebula
Type of Object: Emission nebula and open cluster
Magnitude: 6
Distance: 5,700 ly
Physical Size: Nebula spans at least 58 ly, cluster spans 12 ly
Arc Degree Size in Sky: Nebula extends for 35' plus, cluster is 7' at its greatest width
Other: The Eagle Nebula contains the famous "Pillars of Creation," which is the name given to a widely circulated Hubble Space Telescope photo showing columns of gas where new stars are being born.

NGC Summary by J.L.E. Dreyer, circa 1888

NGC Number: 6611
Mag: 6.0   **Size:** 35'
NGC Description: Cluster, at least 100 bright and faint stars.

Author's Note: Oddly, this NGC description omits anything about the nebulosity associated with this cluster. However, the listing for this entry in *NGC 2000.0* by Sky Publishing does indicate its Type [of object] as a Cluster & Nebula.

Description of M16 using a 4-inch refractor at 48x

This nebula is fairly large and spread out, but not very bright. Overall, it is probably about as bright as the fainter parts of M8. Details within the nebulosity are visible. An open cluster lies on "top" with a section having a high concentration of brighter stars.

Locating Index: Challenging because it is not near any conspicuous stars. If it is positioned near the meridian, I generally locate the much brighter M17 and then move straight up or north until I bump into M16.

Identifying/Observing Index: Slightly challenging. The nebula will be difficult to see in light-polluted skies, so look for the cluster, more easily seen and identifiable at lower magnifications.

ARC SCALE

1° —
3/4° —
1/2° —
25' —
20' —
15' —
10' —
5' —
0' —

TO LOCATE
Refer to
Star Charts
A,
P
& **S**

The nebula part of M16 is not very bright, but in dark skies, it comes alive — much more so than in this picture! Its wings are apparent, even in small telescopes. This photo picked up a "small" meteor.

For easy comparison, all photographs are shown at the same scale, measuring 2.3° x 1.3°.

M17 Omega Nebula ✦ Nebula & Open Cluster

Original Messier Description

from the 1784 edition of *Connaissance des Temps*

Observed 1764. Jun. 3

A streak of light without stars, between five and six minutes in extent, in the shape of a spindle, & more or less like the one in the belt of Andromeda, but with very faint light. There are two telescopic stars nearby & parallel to the Equator. In a good sky, you can see this nebula very well with an ordinary three-&-a-half-foot telescope. Observed again on March 22, 1781.

Location

Constellation: Sagittarius (Sgr)

Year 2000 Coordinates
RA: 18h 20.8m
Dec: –16° 11'

Observation Periods

Evenings 8 p.m. : July to October
Mornings 4 a.m. : March to June

Facts

Name: Omega Nebula, Swan Nebula, Horseshoe Nebula
Type of Object: Emission nebula and open cluster
Magnitude: 7
Distance: 3,000 ly
Physical Size: Nebula spans 40 ly, cluster spans 8 ly
Arc Degree Size in Sky: Nebula 46' x 37', cluster extends to about 9'
Other: Cluster's brightest star has a magnitude of 9.3.

NGC Summary by J.L.E. Dreyer, circa 1888

NGC Number: 6618
Mag: 6.0   **Size:** 46'
NGC Description: A magnificent object!!! Bright, extremely large, extremely irregular figure, hooked like the number "2."

Description of M17 using a 4-inch refractor at 48x

One of my favorite Messier objects. The "bar" is the thickest and brightest part of the nebula and really stands out. The small cluster goes unnoticed because of the magnificence of this nebula.

Locating Index: Slightly challenging because there are no nearby conspicuous stars. If you can see the 4½ magnitude star γ *Scuti*, then you may be able to see the 5th magnitude star that is about 25' away from M17.

Identifying/Observing Index: Easy because it really pops out when you come across it.

TO LOCATE Refer to Star Charts **A, P** & **S**

1° —
ARC SCALE
3/4° —
1/2° —
25' —
20' —
15' —
10' —
5' —
0' —

This is a magnificent object. It is the brightest nebula in the summer sky. The 5th magnitude star that might be visible to your eyes is just above and slightly to the right of the nebula.

For easy comparison, all photographs are shown at the same scale, measuring 2.3° x 1.3°.

M 18 Black Swan ✧ Open Cluster

Original Messier Description

from the 1784 edition of *Connaissance des Temps*

Observed 1764. Jun. 3

Cluster of small stars, a little below the nebula No. 17 listed above; surrounded by a slight nebulosity, this cluster less visible than the preceding one, No. 16. With an ordinary three-&-a-half-foot telescope, this cluster looks like a nebula; but with a better telescope, you only see stars there.

Facts

Name: Black Swan
Type of Object: Open cluster
Magnitude: 6.9
Distance: 3,900 ly
Physical Size: 10 ly at its widest
Arc Degree Size in Sky: 9' at its widest
Other: Contains about 20 stars, the brightest having a magnitude of 8.7. Its age is estimated at 32,000,000 years.

NGC Summary by J.L.E. Dreyer, circa 1888

NGC Number: 6613
Mag: 6.9   **Size:** 9'
NGC Description: Cluster, sparse in stars, very little compressed.

Description of M18 using a 4-inch refractor at 48x

A "weak" cluster that could easily be passed up (like M103) and missed. Pops out better with averted vision because there are about 10 stars below the threshold of direct vision with a small telescope. Relatively compact.

Locating Index: **Slightly challenging** because it is not near any conspicuously bright stars. Find M17 first and "drop" south to locate M18.

Identifying/Observing Index: **Slightly challenging** with a small telescope because it does not "pop out" like many other Messier clusters.

TO LOCATE
Refer to
Star Charts
A,
P
& **S**

ARC SCALE

1° —
3/4° —
1/2° —
25' —
20' —
15' —
10' —
5' —
0' —

M18

N

M17 is at the very top of this picture. M18, comprised of about 20 stars, is the unflattering clump to the left of its designation.

For easy comparison, all photographs are shown at the same scale, measuring 2.3° x 1.3°.

M 19 ✧ Globular Cluster

Original Messier Description

from the 1784 edition of *Connaissance des Temps*

Observed 1764. Jun. 5

Nebula without stars on the parallel of *Antares*, between Scorpius & the right foot of *Ophiuchus*. This nebula is round; you could see it very well with an ordinary three-&-a-half-foot telescope. The nearest known neighboring star is the twenty-eighth of *Ophiuchus*, 6th magnitude, according to Flamsteed. Observed again on March 22, 1781.

NGC Summary by J.L.E. Dreyer, circa 1888

NGC Number: 6273
Mag: 7.2   **Size:** 13.5'
NGC Description: Globular cluster of stars, very bright, large, round, very compressed in the middle, well resolved, contains stars of magnitude 16.

Location

Constellation: Ophiuchus (Oph)
Year 2000 Coordinates
RA: 17h 02.6m
Dec: −26° 16'

Observation Periods

Evenings 8 p.m. : June to September
Mornings 4 a.m. : February to May

Facts

Name: No common name
Type of Object: Globular cluster
Magnitude: 6.8
Distance: 28,000 ly
Physical Size: 110 ly in diameter
Arc Degree Size in Sky: 13½' in diameter
Other: Shows an amount of elongation that is very apparent in pictures. The compactness of this globular is looser than average.

Description of M19 using a 4-inch refractor at 48x

A "nice" globular that appears to be about one-third the size of M4. With averted vision, a few individual stars can be glimpsed near its center, which is bright and somewhat fuzzy.

Locating Index: **Slightly challenging** because it is not near any conspicuous stars. However, it forms the apex of an equilateral triangle using *Antares* and *ε Scorpii* as a base.

Identifying/Observing Index: **Easy** because it is bright for its size so it "pops" when you come across it.

TO LOCATE
Refer to
Star Charts
A,
P
& **S**

M19 has an elongation that is noticeable in this picture, but less so visually.

M19

1° —
ARC SCALE
3/4° —
1/2° —
25' —
20' —
15' —
10' —
5' —
0' —

For easy comparison, all photographs are shown at the same scale, measuring 2.3° x 1.3°.

M20 Trifid Nebula ✧ Nebula & Open Cluster

Original Messier Description

from the 1784 edition of *Connaissance des Temps*

Observed 1764. Jun. 5

Cluster of stars a little above the Ecliptic, between the bow of Sagittarius & the right foot of *Ophiuchus*. Observed again on March 22, 1781.

Location

Constellation: Sagittarius (Sgr)

Year 2000 Coordinates
RA: 18h 02.6m
Dec: −23° 02'

Observation Periods

Evenings 8 p.m. : July to October
Mornings 4 a.m. : March to June

Facts

Name: Trifid Nebula, the Clover
Type of Object: Nebula and open cluster
Magnitude: 8
Distance: 3,500 ly
Physical Size: Nebula measures 29 ly on its "sides" and the cluster spans 29 ly
Arc Degree Size in Sky: Nebula 28' x 28', open cluster extends 28'
Other: Open cluster contains about 60 stars with its brightest shining at magnitude 7.3.

NGC Summary by J.L.E. Dreyer, circa 1888

NGC Number: 6514
Mag: 6.3   **Size:** 29'
NGC Description: A magnificent object!!! Very bright, very large, trifid, double star involved.

Author's Note: The word "trifid" is defined as something that is deeply and narrowly cleft into three parts.

Description of M20 using a 4-inch refractor at 48x

Two circular patches of nebulosity touching one another, each having a fairly bright star near their centers. The brightest parts of these nebulae are fainter than the brightest parts of M8, about a degree south. The large cluster goes mostly unnoticed because the nebula commands your attention.

Locating Index: Fairly easy. First find the easier to spot M8 and then move the telescope one to two eyepiece fields of view north to "bump" into M20 or M21.

Identifying/Observing Index: Easy to spot in dark skies but will prove more troublesome in skies with light pollution. In this case, your only alternative will be to use the picture below to identify the stars in and around the nebulosity.

TO LOCATE
Refer to
Star Charts
A,
P
& **S**

ARC SCALE

1°—
3/4°—
1/2°—
25'—
20'—
15'—
10'—
5'—
0'—

M21

The two touching clouds to the right of center make up M20. The nearby cluster M21 is noted. The clefts of the trifid are visible in the nebula. For M20's position with respect to M8, see the picture for M21.

For easy comparison, all photographs are shown at the same scale, measuring 2.3° x 1.3°.

M21 ✧ Open Cluster

Original Messier Description

from the 1784 edition of *Connaissance des Temps*

Observed 1764. Jun. 5

Star cluster near the preceding one; the known star closest to these two clusters is the eleventh of Sagittarius, seventh magnitude, according to Flamsteed. The stars of these two clusters are of eighth to ninth magnitude, surrounded by nebulosity.

Location

Constellation: Sagittarius (Sgr)

Year 2000 Coordinates
RA: 18h 04.6m
Dec: −22° 30'

Observation Periods

Evenings 8 p.m. : July to October
Mornings 4 a.m. : March to June

Facts

Name: No common name
Type of Object: Open cluster
Magnitude: 5.9
Distance: 4,300 ly
Physical Size: Spans 16 ly
Arc Degree Size in Sky: Extends 13'
Other: Consists of about 70 stars. The brightest star has a magnitude of 7.3. Its age is estimated at a young 4,600,000 years.

NGC Summary by J.L.E. Dreyer, circa 1888

NGC Number: 6531
Mag: 5.9   **Size:** 13'
NGC Description: Cluster, pretty rich in stars, little compressed, contains stars of magnitude 9 to 11.

Description of M21 using a 4-inch refractor at 48x

Not as prominent as other clusters in this catalogue. It would stand out more if it were not in the Milky Way. Relatively small. With direct vision, I can count about 20 stars, but I can see more with averted vision.

Locating Index: Fairly easy as long as you can locate M8 and M20.

Identifying/Observing Index: Easy or **slightly challenging**. Easy if you can see M20 because M21 is only one Moon's width away. Slightly challenging if you cannot see M20 because it does not look too different from many other "clumps" of stars in this area of the Milky Way.

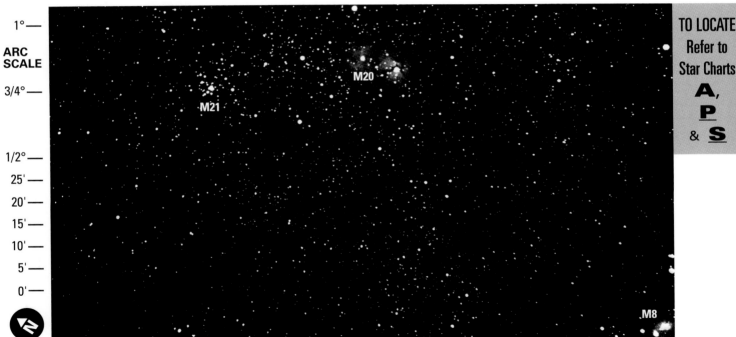

ARC SCALE

1° —
3/4° —
1/2° —
25' —
20' —
15' —
10' —
5' —
0' —

M21

M20

M8

TO LOCATE
Refer to
Star Charts
A,
P
& **S**

The open cluster M21 is ½° to the northeast of M20. It has one bright 7th magnitude star at its center. M8 is visible in the lower right-hand corner.

For easy comparison, all photographs are shown at the same scale, measuring 2.3° x 1.3°.

M22 Great Sagittarius Cluster ✧ Globular Cluster

Original Messier Description

from the 1784 edition of *Connaissance des Temps*

Observed 1764. Jun. 5

Nebula below the Ecliptic, between the head & bow of Sagittarius, near a seventh magnitude star, the twenty-fifth of Sagittarius, according to Flamsteed. This nebula is round, contains no stars, & you can see it very well with an ordinary 3.5-foot telescope. The star λ of Sagittarius was used to determine it. Abraham Ihle, a German, discovered it in 1665 while observing Saturn. Mr. le Gentil observed it in 1747 & had a picture of it printed. *Mémoires de l'Académie, year 1759, page 470.* Observed again on March 22, 1781: it's reported in the English Atlas.*

NGC Summary by J.L.E. Dreyer, circa 1888

NGC Number: 6656
Mag: 5.1 **Size:** 24'
NGC Description: Very remarkable!! Globular cluster of stars, very bright, very large, round, very rich in stars, very much compressed, contains stars of magnitude 11 to 15.

Location

Constellation: Sagittarius (Sgr)
Year 2000 Coordinates
RA: 18h 36.4m
Dec: –23° 54'

Observation Periods

Evenings 8 p.m. : July to October
Mornings 4 a.m. : March to June

Facts

Names: Great Sagittarius Cluster, Crackerjack Cluster
Type of Object: Globular cluster
Magnitude: 5.1
Distance: 10,000 ly
Physical Size: 70 ly in diameter
Arc Degree Size in Sky: 24' in diameter
Other: It is estimated that M22 contains about 70,000 stars. Its compactness is looser than average.

* Author's Note: Johann Abraham Ihle (1627–1699?) was a German post-office official and amateur astronomer who observed sunspots, planets and comets. He was a friend of Kirch and knew Hevelius.

Description of M22 using a 4-inch refractor at 48x

One of my favorites. Beautiful. Appears as many faint stars sprinkled over a large area. The center is the brightest part but overall it has a very even and extended brightness.

Locating Index: Fairly easy because it roughly forms a parallelogram with *Nunki, φ* and *λ Sagittarii*.

Identifying/Observing Index: Easy because it is big and bright, so it "pops" when you come across it.

TO LOCATE
Refer to
Star Charts
**A,
P**
& **S**

ARC SCALE: 1°, 3/4°, 1/2°, 25', 20', 15', 10', 5', 0'

N

M22 is magnificent, even in small diameter telescopes. In angular size, it is the fourth largest in the sky, and appears much bigger than M13.

For easy comparison, all photographs are shown at the same scale, measuring 2.3° x 1.3°.

M23 ✦ Open Cluster

Original Messier Description

from the 1784 edition of *Connaissance des Temps*

Observed 1764. Jun. 20

A star cluster between the tip of the bow of Sagittarius & the right foot of *Ophiuchus*, very close to the 65th star of *Ophiuchus*, according to Flamsteed. The stars of this cluster are very close to each other. Its position determined from μ of Sagittarius.

Location

Constellation: Sagittarius (Sgr)

Year 2000 Coordinates
RA: 17h 56.8m
Dec: −19° 01'

Observation Periods

Evenings 8 p.m. : July to October
Mornings 4 a.m. : March to June

Facts

Name: No common name
Type of Object: Open cluster
Magnitude: 5½
Distance: 1,400 ly
Physical Size: Spans 11 ly
Arc Degree Size in Sky: Extends 27'
Other: Contains about 150 stars, the brightest shining at magnitude 9.2. Its age is estimated at 220,000,000 years.

NGC Summary by J.L.E. Dreyer, circa 1888

NGC Number: 6494
Mag: 5.5   **Size:** 27'
NGC Description: Cluster, bright, very large, pretty rich in stars, little compressed, contains stars of 10th magnitude and fainter.

Description of M23 using a 4-inch refractor at 48x

Pretty cluster. A random sprinkle of similarly colored and like-magnitude stars. There is one star that is much brighter than the rest. What is surprising is the degree to which this cluster stands out despite being in a dense part of the Milky Way.

Locating Index: Challenging because it is not near any conspicuous stars. It does, however, form the apex of a shallow isosceles triangle with λ *Sagittarii* and *Sabik* in Ophiuchus.

Identifying/Observing Index: Easy because it is large and visually striking.

TO LOCATE
Refer to
Star Charts
A,
P
& **S**

ARC
SCALE

1° —
3/4° —
1/2° —
25' —
20' —
15' —
10' —
5' —
0' —

M23 is a rich open cluster that is in a thick patch of the Milky Way.

For easy comparison, all photographs are shown at the same scale, measuring 2.3° x 1.3°.

M24 ✧ Thick Patch of the Milky Way

Original Messier Description

from the 1784 edition of *Connaissance des Temps*

Observed 1764. Jun. 20

Cluster on the parallel of the preceding one & near the tip of the bow of Sagittarius, in the Milky Way: a large nebulosity in which there are many stars of different sizes: the light that spreads through this cluster is divided into several areas. It's the center of this cluster that has been determined.

Location

Constellation: Sagittarius (Sgr)
Year 2000 Coordinates
RA: 18h 16.9m
Dec: –18° 29'

Observation Periods

Evenings 8 p.m. : July to October
Mornings 4 a.m. : March to June

Facts

Name: Small Sagittarius Star Cloud
Type of Object: Thick patch of the Milky Way consisting of millions to perhaps a billion or so stars "clumped" together in the same "area" or direction
Magnitude: Estimates vary from 2.5 to 4.6
Distance: Not applicable because the stars that make up this patch are at different distances
Physical Size: Not applicable
Arc Degree Size in Sky: 90' x 60'
Other: The stars that make up this patch reside in an inner arm of our galaxy known as the Sagittarius arm.

NGC Summary by J.L.E. Dreyer, circa 1888

NGC Number: Not listed as an NGC object because Milky Way patches do not fall into the category of objects listed and described in the catalogue.

Description of M24 using a 4-inch refractor at 48x

An incredible, rich patch of the Milky Way. An amazing number of "things" are going on within this large area. I can see dark holes and dark areas. Using averted vision, I notice two clusters, one small and compact (NGC 6603), the other looser and near one of the holes.

Locating Index: Slightly challenging because it is not near any conspicuous stars. It starts about one degree north of the 4th magnitude star *μ Sagittarii*.

Identifying/Observing Index: Fairly easy in dark skies because it stands out, but to see it, you must use a lower magnification that provides a wide field of view. This "object" is probably more challenging, if not impossible to see in light-polluted skies.

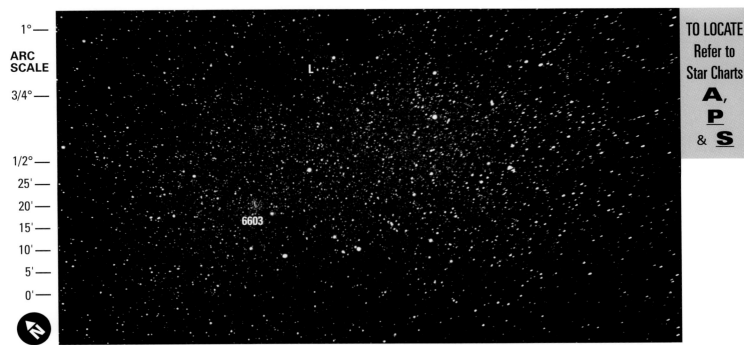

ARC SCALE

1°
3/4°
1/2°
25'
20'
15'
10'
5'
0'

6603

L

TO LOCATE
Refer to
Star Charts
A,
P
& **S**

M24 practically spans this whole picture. The brightest part is to the right of center. The small compact cluster NGC 6603 is noted. The location of the looser cluster is indicated with an L.

For easy comparison, all photographs are shown at the same scale, measuring 2.3° x 1.3°.

M25 ✦ Open Cluster

Original Messier Description

from the 1784 edition of *Connaissance des Temps*

Observed 1764. Jun. 20

Cluster of small stars in the vicinity of the two previous clusters between the head & the tip of the bow of Sagittarius; the known star closest to this cluster is the star 21 of Sagittarius, 6th magnitude according to Flamsteed. The stars of this cluster are hard to see with an ordinary three-foot telescope; no nebulosity is visible. Its position was known from the star µ of Sagittarius.

Location

Constellation: Sagittarius (Sgr)
Year 2000 Coordinates
RA: 18h 31.6m
Dec: −19° 15'

Observation Periods

Evenings 8 p.m. : July to October
Mornings 4 a.m. : March to June

Facts

Name: No common name
Type of Object: Open cluster
Magnitude: 4.6
Distance: 2,000 ly
Physical Size: Spans 19 ly
Arc Degree Size in Sky: Extends 32'
Other: Contains about 30 stars, the brightest shining at magnitude 6.7. Its age is estimated at 89,000,000 years.

IC Summary by J.L.E. Dreyer, circa 1908

IC Number: 4725
Mag: 4.6   **Size:** 32'
IC Description: Cluster, pretty compressed.

Description of M25 using a 4-inch refractor at 48x

Overall, its shape reminds me of a cross. Most of the stars appear inside a trapezoid roughly formed by four stars. This is a nice open cluster and would be a good candidate as representative of what's "typical."

Locating Index: Slightly challenging because it is not near any conspicuous stars, but it forms the apex of an isosceles triangle with λ *Sagittarii* and γ *Scuti*.

Identifying/Observing Index: Easy as long as you are using a lower magnification to provide a wide field of view, otherwise you could pass it up.

ARC SCALE

1° —
3/4° —
1/2° —
25' —
20' —
15' —
10' —
5' —
0' —

N

TO LOCATE
Refer to
Star Charts
A,
P
& **S**

M25 is a good example of a "typical" open cluster. It is about 4° due east of M24.

For easy comparison, all photographs are shown at the same scale, measuring 2.3° x 1.3°.

M26 ✧ Open Cluster

Original Messier Description

from the 1784 edition of *Connaissance des Temps*

Observed 1764. Jun. 20

A star cluster near the stars *n* & *o* in Antinoüs in between which there is one which has more light: they can't be distinguished with a three-foot telescope, a good instrument must be used. This cluster contains no nebulosity.

Author's Notes: The constellation Aquila has replaced the constellation Antinoüs. Antinoüs, a youth, was honored with a place in the sky (circa A.D. 132) after sacrificing his life for the Roman Emperor Hadrian believing that it would prolong the life of his master.
   The stars *n* & *o* are respectively, the stars *ε* & *δ Scuti* and are indicated at the top of Chart S.

NGC Summary by J.L.E. Dreyer, circa 1888

NGC Number: 6694
Mag: 8.0 **Size:** 15'
NGC Description: Cluster, considerably large, pretty rich in stars, pretty compressed, contains stars of magnitude 12 to 16.

Location

Constellation: Scutum (Sct)
Year 2000 Coordinates
RA: 18h 45.2m
Dec: −9° 24'

Observation Periods

Evenings 8 p.m. : July to October
Mornings 4 a.m. : March to June

Facts

Name: No common name
Type of Object: Open cluster
Magnitude: 8.0
Distance: 4,900 ly
Physical Size: Spans 21 ly
Arc Degree Size in Sky: Extends 15'
Other: Contains about 30 stars, the brightest shining at magnitude 10.3. Its age is estimated at 89,000,000 years.

Description of M26 using a 4-inch refractor at 48x

More roundish in shape than others, fairly small and compact. The stars are not as bright as in other Messier clusters. It also appears as though there is some nebulosity, but star charts indicate none. Could this be an optical illusion resulting from the placement and glow of these stars?

Locating Index: Fairly easy to **slightly challenging**. It is within three degrees of the 4th magnitude star *a Scuti*, the brightest star in Scutum.

Identifying/Observing Index: Slightly challenging even in dark skies because it is not big and bright, but more subtle. Will prove challenging in light-polluted skies.

TO LOCATE
Refer to
Star Charts
A,
P, R,
& S

M26 is one of the more compact open clusters in the Messier catalogue.

For easy comparison, all photographs are shown at the same scale, measuring 2.3° x 1.3°.

M27 Dumbbell Nebula ✦ Planetary Nebula

Original Messier Description

from the 1784 edition of *Connaissance des Temps*

Observed 1764. July 12

Nebula without stars, discovered
in Vulpecula, between the two front
paws & very close to the star 14 of this
constellation, 5th magnitude according
to Flamsteed. You can see it well with
an ordinary 3-&-a-half-foot telescope.
It appears as an oval & contains no stars.
Mr. Messier reported its position on
the Chart of the Comet of 1779, which
will be printed for the Acad.'s volume
of the same year. Observed again on
January 31, 1781.

NGC Summary by J.L.E. Dreyer, circa 1888

NGC Number: 6853
Mag: 8.1 **Size:** 15.2'
NGC Description: A magnificent object!!!
Very bright, very large, binuclear, irregu-
larly extended like a dumbbell.

Location

Constellation: Vulpecula (Vul)

Year 2000 Coordinates
RA: 19h 59.6m
Dec: +22° 43'

Observation Periods

Evenings 8 p.m. : June to December
Mornings 4 a.m. : February to August

Facts

Name: Dumbbell Nebula
Type of Object: Planetary nebula
Magnitude: 8
Distance: Estimates range from 815 ly
to 3,500 ly
Physical Size: Spans 1.9 ly for the
closest distance estimate and 8 ly
for the farthest
Arc Degree Size in Sky: 8' x 4'
Other: The central star, that is, the star
that created this shedded atmosphere,
has a magnitude of 13.9. The age of
the nebula is estimated at 20,000 years.

Description of M27 using a 4-inch refractor at 48x

Incredibly bright (I am observing in a good, dark sky). It appears roundish and looks pretty evenly illuminated. However, with averted vision, I can see some lighter and darker areas. I can just make out diametrically opposed darker areas, which give rise to the dumbbell shape.

Locating Index: Slightly challenging because it is not near any conspicuous stars. To find it, triangulate with *Albireo* and the brighter stars of Sagitta and Vulpecula.

Identifying/Observing Index: Easy in dark skies because it is so bright. It will be **slightly challenging** in many light-polluted skies.

ARC SCALE

1° —
3/4° —
1/2° —
25' —
20' —
15' —
10' —
5' —
0' —

TO LOCATE Refer to Star Charts **A, P & R**

In dark skies, the Dumbbell is an absolutely spectacular sight, even in smaller diameter telescopes. Note the interesting string of fainter stars leading to a brighter one 25' due north of the nebula.

For easy comparison, all photographs are shown at the same scale, measuring 2.3° x 1.3°.

M28 ✧ Globular Cluster

Original Messier Description

from the 1784 edition of *Connaissance des Temps*

Observed 1764. July 29

Nebula discovered in the higher part of the bow of Sagittarius about one degree away from the star λ & not far from the beautiful nebula which is between the head & the bow. It contains no stars; it's round, it can only be seen with great difficulty through an ordinary 3½-foot telescope. Its position was determined from λ of Sagittarius. Observed again on March 20, 1781.

NGC Summary by J.L.E. Dreyer, circa 1888

NGC Number: 6626
Mag: 6.9   **Size:** 11.2'
NGC Description: Remarkable! Globular cluster of stars, very bright, large, round, gradually becomes extremely compressed in the middle, well resolved, contains stars of magnitude 14 to 16.

Location

Constellation: Sagittarius (Sgr)
Year 2000 Coordinates
RA: 18h 24.5m
Dec: −24° 52'

Observation Periods

Evenings 8 p.m. : July to October
Mornings 4 a.m. : March to June

Facts

Name: No common name
Type of Object: Globular cluster
Magnitude: 6.8
Distance: 19,000 ly
Physical Size: 62 ly in diameter
Arc Degree Size in Sky: 11.2' in diameter
Other: The compactness of the stars in this globular cluster is tighter than average.

Description of M28 using a 4-inch refractor at 48x

Very bright and pretty concentrated center. On a night of poor seeing or in light-polluted skies, it may only seem starlike. Its basic appearance is of a bright center surrounded by a halo.

Locating Index: Easy because it is only 1 degree from the 3rd magnitude handle star *Kaus Borealis* (λ) in Sagittarius.

Identifying/Observing Index: Easy in dark skies but more **challenging** in light-polluted skies. Under "poor" skies, look for a fuzzy star.

TO LOCATE Refer to Star Charts **A, P & S**

ARC SCALE

1°
3/4°
1/2°
25'
20'
15'
10'
5'
0'

M28 is in a thick part of the Milky Way. *Kaus Borealis* is the bright star to the upper left.

For easy comparison, all photographs are shown at the same scale, measuring 2.3° x 1.3°.

M29 ✧ Open Cluster

Original Messier Description

from the 1784 edition of *Connaissance des Temps*

Observed 1764. July 29

Cluster of seven or eight very small stars which are below γ of Cygnus, that looks like a nebula with an ordinary three-&-a-half-foot telescope. Its position determined by γ of Cygnus. This cluster is reported on the Chart of the Comet of 1779.

Location

Constellation: Cygnus (Cyg)

Year 2000 Coordinates
RA: 20h 23.9m
Dec: +38° 32'

Observation Periods

Evenings 8 p.m. : June to January
Mornings 4 a.m. : February to September

Facts

Name: No common name
Type of Object: Open cluster
Magnitude: 6.6
Distance: 4,000 ly
Physical Size: Spans 8 ly
Arc Degree Size in Sky: Extends 7'
Other: Contains about 50 stars, the brightest shining at magnitude 8.6. Its age is estimated at 10,000,000 years.

NGC Summary by J.L.E. Dreyer, circa 1888

NGC Number: 6913
Mag: 6.6 **Size:** 7'
NGC Description: Cluster, spare in stars, little compressed, stars bright and faint.

Description of M29 using a 4-inch refractor at 48x

Visually not very impressive, but it does contain interesting subtleties. Basically, there are 6 bright stars with 4 of them roughly forming the shape of the Great Square of Pegasus. This cluster "pops" more with averted vision. There appears to be a dark area surrounding the cluster.

Locating Index: Fairly easy because it is within 2 degrees of the center star, *γ Cygni*, that helps form the Northern Cross or Cygnus.

Identifying/Observing Index: Challenging because it does not stand out, but blends in with the surrounding Milky Way stars.

TO LOCATE
Refer to Star Charts
A, Q & R

You could easily pass by M29 because it blends in so well with the surrounding Milky Way. It is centered in this picture. Visually, there appears to be a dark patch surrounding the cluster, as well as one adjacent to it.

For easy comparison, all photographs are shown at the same scale, measuring 2.3° x 1.3°.

M30 ✧ Globular Cluster

Original Messier Description

from the 1784 edition of *Connaissance des Temps*

Observed 1764. Aug. 3

Nebula discovered below the tail of Capricornus, very close to the star 41 of this constellation, 6th magnitude according to Flamsteed. It's hard to see with an ordinary 3½-foot telescope. It's round, & contains no stars; its position determined by ζ of Capricornus. Mr. Messier reported it on the Chart of the Comet of 1759. *Mém. Acad. 1760, pl. II.*

Location

Constellation: Capricornus (Cap)

Year 2000 Coordinates

RA: 21h 40.4m

Dec: −23° 11'

Observation Periods

Evenings 8 p.m. : August to December

Mornings 4 a.m. : May to August

Facts

Name: No common name

Type of Object: Globular cluster

Magnitude: 7.2

Distance: 26,000 ly

Physical Size: 83 ly in diameter

Arc Degree Size in Sky: 11' in diameter

Other: Compactness of its stars is slightly tighter than average. See "Globular Cluster" in the Glossary for more information on compactness.

NGC Summary by J.L.E. Dreyer, circa 1888

NGC Number: 7099

Mag: 7.5   **Size:** 11'

NGC Description: Remarkable! Globular cluster of stars, bright, large, little extended, gradually becomes pretty much brighter in the middle, contains stars of magnitude 12 to 16.

Description of M30 using a 4-inch refractor at 48x

Appears to be about the same brightness as a star 25' away. Center is more extended and surrounded by a fainter but evenly lit ring. Either M30 or galaxy M74 is often the one object that cannot be observed during a Messier Marathon because they get lost in the Sun's glow.

Locating Index: Slightly challenging but it is not far from the 3rd and 4th magnitude stars *36* and *ζ Capricorni* which can be used as guides.

Identifying/Observing Index: Very bright and thus **easy** in dark skies but it fades in light-polluted skies to become **challenging**.

TO LOCATE Refer to Star Charts **A, D & R**

M30 is the early morning object that sometimes "gets away" from being observed during Messier Marathons. It is 20' from the 5th magnitude star *41 Capricorni*.

For easy comparison, all photographs are shown at the same scale, measuring 2.3° x 1.3°.

M31 Andromeda Galaxy ✧ Galaxy

Original Messier Description

from the 1784 edition of *Connaissance des Temps*

Observed 1764. Aug. 3

The beautiful nebula in the belt of Androm-eda, shaped like a spindle. Mr. Messier has investigated it with different instruments, & has found no stars in it. It resembles two cones or pyramids of light, with vertically opposed bases, whose axis runs in a north-west-to-southeast direction. The two points of light or peaks were about 40 minutes of arc away from each other; the pyramids' common base of 15 minutes. This nebula was discovered in 1612 by Simon *Marius*, & afterwards observed by different Astron-omers. Mr. le Gentil has put a drawing of it in the Mémoires de l'Académie of 1759, *page 453*. It's reported on the English *Atlas*.*

NGC Summary by J.L.E. Dreyer, circa 1888

NGC Number: 224
Mag: 3.5 **Size:** 178'
NGC Description: A magnificent object!!! Most extremely bright, extremely large, very much extended. Known as Andromeda.

Location

Constellation: Andromeda (And)

Year 2000 Coordinates of the Core
RA: 0h 42.7m
Dec: +41° 16'

Observation Periods

Evenings 8 p.m. : August to March
Mornings 4 a.m. : April to November

Facts

Name: Andromeda Galaxy
Type of Object: Spiral galaxy
Magnitude: 3.5
Distance: 2,400,000 ly
Physical Size: 120,000 ly in diameter
Arc Degree Size in Sky: 2.97° x 1.05° (178' x 63')
Other: Andromeda is about 50% longer than our galaxy and possibly contains more than 300,000,000,000 stars. Its shape is classified as Sb, meaning that it has a round nucleus with medium wound arms. Evidence suggests that it may collide with our Milky Way Galaxy in about 4,000,000,000 years. It is a member of our Local Group of galaxies.

* Author's Note: Simon Marius (1573–1624) studied under Tycho Brahe.

Description of M31 using a 4-inch refractor at 48x

This galaxy looks like a giant elliptical galaxy instead of a spiral, because its arms are not noticeable. The center of the large glow appears slightly starlike, with an extended brightness surrounding it that then fades off. It fills about one-half of the eyepiece field of view (Plössl design).

Locating Index: Fairly easy. To locate, use the stars μ and β *Andromedae*. The galaxy is in line with these stars and as far away from μ as μ is from β.

Identifying/Observing Index: Fairly easy, even in many light-polluted skies because it is just visible to the naked eye. Not as bright in a telescope as you might think and you only really see its core.

ARC SCALE

1° —
3/4° —
1/2° —
25' —
20' —
15' —
10' —
5' —
0' —

N

TO LOCATE
Refer to
Star Charts
**B,
C**
& **E**

This picture is a fairly accurate small-telescope representation of the Andromeda Galaxy because for the most part, all you see is the galaxy's core. Also look at the picture for M32.

For easy comparison, all photographs are shown at the same scale, measuring 2.3° x 1.3°.

M32 ✧ Galaxy

Original Messier Description

from the 1784 edition of *Connaissance des Temps*

Observed 1764. Aug. 3

Small nebula without stars, below & a few minutes away from the one in the belt of Andromeda; this little nebula is round, with much fainter light than the one in the belt. Mr. le Gentil discovered it on October 29, 1749. Mr. Messier saw it for the first time in 1757, & found no change in it.

NGC Summary by J.L.E. Dreyer, circa 1888

NGC Number: 221
Mag: 8.2   **Size:** 7.6'
NGC Description: Remarkable! *Very* bright, large, round, pretty suddenly becomes much brighter in the middle to the nucleus.

Location

Constellation: Andromeda (And)

Year 2000 Coordinates
RA: 0h 42.7m
Dec: +40° 52'

Observation Periods

Evenings 8 p.m. : August to March
Mornings 4 a.m. : April to November

Facts

Name: No common name
Type of Object: Dwarf elliptical galaxy
Magnitude: 8.2
Distance: 2,400,000 ly
Physical Size: Spans at least 5,600 ly
Arc Degree Size in Sky: 8' x 6'
Other: A companion galaxy, gravationally bound to the Andromeda Galaxy. Its shape is classified as E2 which means that it is slightly elongated. Elliptical galaxies are classified from E0 which are perfectly round, to E7 which are very elongated. See illustration on page 340.

Description of M32 using a 4-inch refractor at 48x

It is about 20' from the core of Andromeda. It appears starlike with a little luminosity around it, similar to some globular clusters.

Locating Index: Easy, because it is about a Moon's diameter from the center or core of the Andromeda Galaxy (M31).

Identifying/Observing Index: Easy, even in light-polluted skies. It is bright and appears as a fuzzy ball.

TO LOCATE
Refer to
Star Charts
**B,
C**
& **E**

ARC SCALE

1° —
3/4° —
1/2° —
25' —
20' —
15' —
10' —
5' —
0' —

M110

M32

An "enhanced" version of the picture on page 169 to bring out the arms of Andromeda. Although M32 appears at the edge of the larger galaxy, through a small telescope, it looks detached from the core because the arms are not this apparent. For easy comparison, all photographs are shown at the same scale, measuring 2.3° x 1.3°.

M33 Pinwheel Galaxy ✦ Galaxy

Original Messier Description

from the 1784 edition of *Connaissance des Temps*

Observed 1764. Aug. 25

Nebula discovered between the head of the northern Fish & Triangulum, not far from a 6th magnitude star: this nebula has a whitish light of almost even density, but a little brighter at two-thirds of its diameter, & contains no stars. It's hard to see with an ordinary one-foot telescope. Its position determined by comparing it to α of Triangulum. Observed again on Sept. 27, 1780.

NGC Summary by J.L.E. Dreyer, circa 1888

NGC Number: 598
Mag: 5.7 **Size:** 62'
NGC Description: Remarkable! Extremely bright, extremely large, round, very gradually becomes brighter in the middle to the nucleus.

Location

Constellation: Triangulum (Tri)
Year 2000 Coordinates
RA: 1h 33.9m
Dec: +30° 39'

Observation Periods

Evenings 8 p.m. : September to March
Mornings 4 a.m. : May to November

Facts

Name: Pinwheel Galaxy, Triangulum Galaxy
Type of Object: Spiral galaxy
Magnitude: 5.7
Distance: 2,200,000 ly
Physical Size: 40,000 ly in diameter
Arc Degree Size in Sky: 62' x 39'
Other: Very large but very faint. You may find it easier to see this galaxy with a pair of binoculars than through a telescope. The shape of this galaxy is classified as Sc, which means that it has a round nucleus with loosely wound arms. It is a member of our Local Group of galaxies.

Description of M33 using a 4-inch refractor at 48x

I have always found this spiral galaxy difficult to see. It is large, face on and very diffused. Its "nucleus" is about one-half the Moon's diameter, housing a smaller brighter core. Around this nucleus is an even fainter area and beyond, where the arms are located, it gets even fainter.

Locating Index: **Fairly easy** to find its position, but more difficult to see it. To locate, use the stars μ and β *Andromedae*. The galaxy is in line with these stars and almost *twice* the distance from β as β is from μ.

Identifying/Observing Index: Can be **difficult** because it is very faint, including the core. It needs dark skies, and easily gets washed out in light-polluted skies.

M33

ARC SCALE

1°
3/4°
1/2°
25'
20'
15'
10'
5'
0'

TO LOCATE
Refer to
Star Charts
A,
D
& **F**

M33 is much brighter in this picture than visually through a small telescope. A few of my friends report seeing it with their naked eyes. Yeah, only in my dreams.

For easy comparison, all photographs are shown at the same scale, measuring 2.3° x 1.3°.

M34 ✧ Open Cluster

Original Messier Description

from the 1784 edition of *Connaissance des Temps*

Observed 1764. Aug. 3

Cluster of small stars between the head of Medusa & the left foot of Andromeda, almost below the parallel of γ. With an ordinary 3-foot telescope you can distinguish the stars. Its position was determined by β of the head of Medusa.

Location

Constellation: Perseus (Per)

Year 2000 Coordinates
RA: 2h 42.0m
Dec: +42° 47'

Observation Periods

Evenings 8 p.m. : September to April
Mornings 4 a.m. : May to December

Facts

Name: No common name
Type of Object: Open cluster
Magnitude: 5.2
Distance: 1,400 ly
Physical Size: Spans 14 ly
Arc Degree Size in Sky: Extends 35'
Other: Contains about 60 stars, some red in color. The brightest shines at magnitude 7.3. Its age is estimated at 190,000,000 years.

NGC Summary by J.L.E. Dreyer, circa 1888

NGC Number: 1039
Mag: 5.2   **Size:** 35'
NGC Description: Cluster, bright, very large, little compressed, contains scattered stars of 9th magnitude.

Description of M34 using a 4-inch refractor at 48x

Nice, large, bright open cluster that "pops" out at you when you come across it. It fills about the space of a Full Moon. A number of its brighter stars create a circle, and within, near the center, lies the greatest concentration of stars.

Locating Index: Fairly easy because it forms the apex of a shallow isosceles triangle with *Algol* and *γ Andromedae*.

Identifying/Observing Index: Easy because it is big and bright, so it "pops" when you come across it.

TO LOCATE
Refer to
Star Charts
**B,
C, E**
& **G**

ARC SCALE

1° —
3/4° —
1/2° —
25' —
20' —
15' —
10' —
5' —
0' —

N

M34 is a pretty large sprinkle of relatively bright stars.

For easy comparison, all photographs are shown at the same scale, measuring 2.3° x 1.3°.

M35 ✦ Open Cluster

Original Messier Description

from the 1784 edition of *Connaissance des Temps*

Observed 1764. Aug. 30

Cluster of very small stars, near the left foot of Castor, not far from the stars μ & η of that constellation. Mr. Messier reported its position on the Chart of the Comet of 1770. *Mém. Acad. 1771, pl. VII.* Reported in the English *Atlas*.

Location

Constellation: Gemini (Gem)

Year 2000 Coordinates
RA: 6h 08.9m
Dec: +24° 20'

Observation Periods

Evenings 8 p.m. : November to May
Mornings 4 a.m. : August to January

Facts

Name: No common name
Type of Object: Open cluster
Magnitude: 5.1
Distance: 2,800 ly
Physical Size: Spans 23 ly
Arc Degree Size in Sky: Extends 28'
Other: Contains about 200 stars, the brightest shining at magnitude 8.2. Its age is estimated at 110,000,000 years.

NGC Summary by J.L.E. Dreyer, circa 1888

NGC Number: 2168
Mag: 5.1   **Size:** 28'
NGC Description: Cluster, very large, considerably rich in stars, pretty compressed, contains stars of magnitude 9 to 16.

Description of M35 using a 4-inch refractor at 48x

Covers an area of about a Moon's diameter. There appear to be about 100 visible stars that fit into the shape of an oval, and many more fainter ones can be glimpsed with averted vision. The small cluster NGC 2158 can be discerned more easily with averted vision.

Locating Index: Easy because it forms the right angle of a small triangle with the last two 3rd and 4th magnitude "feet" stars (η & ι) in Gemini.

Identifying/Observing Index: Easy because it is large, pretty and relatively bright when you come across it.

TO LOCATE
Refer to
Star Charts
B,
H
& **J**

1°—
ARC SCALE
3/4°—
1/2°—
25'—
20'—
15'—
10'—
5'—
0'—

2158

This cluster is a favorite that is easy to find. As a bonus, it sports the small NGC 2158 cluster within 20' of its center.

For easy comparison, all photographs are shown at the same scale, measuring 2.3° x 1.3°.

M36 ✧ Open Cluster

Original Messier Description

from the 1784 edition of *Connaissance des Temps*

Observed 1764. Sept. 2

Cluster of stars in Auriga, near the star φ. With an ordinary telescope of 3 & a half feet it's difficult to distinguish the stars. The cluster contains no nebulosity. Its position is determined from φ.

Location

Constellation: Auriga (Aur)

Year 2000 Coordinates
RA: 5h 36.1m
Dec: +34° 08'

Observation Periods

Evenings 8 p.m. : November to May
Mornings 4 a.m. : July to January

Facts

Name: No common name
Type of Object: Open cluster
Magnitude: 6.0
Distance: 3,700 ly
Physical Size: Spans 13 ly
Arc Degree Size in Sky: Extends for 12'
Other: Consists of about 60 stars with the brightest shining at magnitude 8.9. Its age is estimated at 25,000,000 years.

NGC Summary by J.L.E. Dreyer, circa 1888

NGC Number: 1960
Mag: 6.0 **Size:** 12'
NGC Description: Cluster, bright, very large, very rich in stars, little compressed, contains stars of magnitude 9 to 11 scattered.

Description of M36 using a 4-inch refractor at 48x

There are about a dozen "bright" stars that define this cluster and some fainter ones. Overall, it's about the same size as M37 but with brighter stars. It is shaped somewhat like a starfish with a missing arm.

Locating Index: **Easy** because it forms the apex of a shallow isosceles triangle with one side of Auriga's pentagon.

Identifying/Observing Index: **Easy** because it "pops" out at you when you come across it.

TO LOCATE Refer to Star Charts **B**, **G** & **H**

The trio of clusters M36 to M38 is easy to find and makes for wonderful viewing. Try them with binoculars too.

For easy comparison, all photographs are shown at the same scale, measuring 2.3° x 1.3°.

179

M37 ✧ Open Cluster

Original Messier Description

from the 1784 edition of *Connaissance des Temps*

Observed 1764. Sept. 2

Cluster of small stars, not far from the preceding one, on the parallel of χ of Auriga; the stars are smaller, closer together & contain some nebulosity. With an ordinary telescope of 3 & a half feet, it's hard to see the stars. This cluster is reported on the Chart of the Comet of 1771, *Mém. Acad. 1777.*

NGC Summary by J.L.E. Dreyer, circa 1888

NGC Number: 2099
Mag: 5.6   **Size:** 24'
NGC Description: Cluster, rich in stars, pretty compressed in the middle, stars bright and faint.

Location

Constellation: Auriga (Aur)
Year 2000 Coordinates
RA: 5h 52.4m
Dec: +32° 33'

Observation Periods

Evenings 8 p.m. : November to May
Mornings 4 a.m. : July to January

Facts

Name: No common name
Type of Object: Open cluster
Magnitude: 5.6
Distance: 4,200 ly
Physical Size: Spans 29 ly
Arc Degree Size in Sky: Extends 24'
Other: Contains about 150 stars with the brightest shining at magnitude 9.2. Its age is estimated at 300,000,000 years.

Description of M37 using a 4-inch refractor at 48x

One of my favorites because it boasts a brighter red-colored star set within a mist of fainter stars. The color of the red star is enhanced by larger telescopes. The fainter stars roughly form the shape of a "V."

Locating Index: Fairly easy. Almost "across" and the same distance away from the θ/β side of Auriga as M36, but a little closer to θ *Aurigae*.

Identifying/Observing Index: Fairly easy in dark skies, but will prove more challenging in light-polluted skies because its mist of fainter stars will not be as visible.

ARC SCALE

1°
3/4°
1/2°
25'
20'
15'
10'
5'
0'

TO LOCATE
Refer to
Star Charts
B,
G
& **H**

The bright red star is close to the middle of M37.

For easy comparison, all photographs are shown at the same scale, measuring 2.3° x 1.3°.

M38 ✧ Open Cluster

Original Messier Description

from the 1784 edition of *Connaissance des Temps*

Observed 1764. Sept. 25

Cluster of small stars in Auriga, near the star σ, not far from the two preceding clusters; this one has a square shape & contains no nebulosity, if examined carefully with a good telescope. It could extend to 15 minutes of arc.

Location

Constellation: Auriga (Aur)
Year 2000 Coordinates
RA: 5h 28.7m
Dec: +35° 50'

Observation Periods

Evenings 8 p.m. : November to May
Mornings 4 a.m. : July to January

Facts

Name: No common name
Type of Object: Open cluster
Magnitude: 6.4
Distance: Estimates range from 2,800 ly to 4,600 ly
Physical Size: Spans 17 ly for the closest distance estimate and 28 ly for the farthest
Arc Degree Size in Sky: Extends 21'
Other: Contains about 100 stars with the brightest shining at magnitude 9.5. Its age is estimated at 220,000,000 years.

NGC Summary by J.L.E. Dreyer, circa 1888

NGC Number: 1912
Mag: 6.4   **Size:** 21'
NGC Description: Cluster, bright, very large, very rich in stars, irregular figure, stars bright and faint.

Description of M38 using a 4-inch refractor at 48x

Composed of a number of stars that are fairly bright. You can see several shapes in this menagerie. There is a "bar" of stars in the middle which helps to shape the letter "A" with the other stars around it.

Locating Index: Fairly easy because it forms the apex of an isosceles triangle with a side of Auriga's pentagon (stars θ and β).

Identifying/Observing Index: Easy because it "pops" when you come across it.

TO LOCATE
Refer to Star Charts
B,
G
& **H**

About 30' from M38 is the fainter cluster NGC 1907. Both are visible with binoculars.

For easy comparison, all photographs are shown at the same scale, measuring 2.3° x 1.3°.

M39 ✦ Open Cluster

Original Messier Description

from the 1784 edition of *Connaissance des Temps*

Observed 1764. Oct. 24

Cluster of stars near the tail of the Swan; you can see them with an ordinary telescope of 3 & a half feet.

Location

Constellation: Cygnus (Cyg)

Year 2000 Coordinates
RA: 21h 32.2m
Dec: +48° 26'

Observation Periods

Evenings 8 p.m. : June to February
Mornings 4 a.m. : February to October

Facts

Name: No common name
Type of Object: Open cluster
Magnitude: 4.6
Distance: 800 ly
Physical Size: Spans 7½ ly
Arc Degree Size in Sky: Extends 32'
Other: About 30 stars with the brightest shining at magnitude 6.8. Its age is estimated at 270,000,000 years.

NGC Summary by J.L.E. Dreyer, circa 1888

NGC Number: 7092
Mag: 4.6   **Size:** 32'
NGC Description: Cluster, very large, very sparse in stars, very little compressed, contains stars of magnitude 7 to 10.

Description of M39 using a 4-inch refractor at 48x

Large cluster with about a dozen bright members, some around magnitude 6/7. It's about a Moon's diameter in size.

Locating Index: Challenging because it is not near any conspicuously bright stars. To get close to its position, use several of the 4th magnitude stars north of *Deneb* as stepping stones. M39 also forms the right angle of a triangle with *Deneb* and *α Cephei*.

Identifying/Observing Index: Easy as long as you are using low magnification since it is big and bright.

TO LOCATE
Refer to
Star Charts
A,
C
& **Q**

ARC SCALE

1°—
3/4°—
1/2°—
25'—
20'—
15'—
10'—
5'—
0'—

N

M39 is big, with stars that are spread out, so it is also a nice object for binoculars.

For easy comparison, all photographs are shown at the same scale, measuring 2.3° x 1.3°.

M40 ✦ Double Star

Original Messier Description

from the 1784 edition of *Connaissance des Temps*

Observed 1764. Oct. 24

Two stars very close together & very small, located at the base of the tail of Ursa Major: They are difficult to distinguish with an ordinary 6-foot telescope. It was while searching for the nebula located above the tail of Ursa Major, which is reported in the book about the Constellation Figures and which, in 1660, must have been at 183^d 32' 41" right ascension & 60^d 20' 33" northern declination, and that Mr. Messier wasn't able to see, that he observed these two stars.

Author's Notes: I once erronously thought that M40 was a mistake made by Messier, that it was an object he thought showed some nebulosity. However, it is clear from this description that he was searching for a nebula reported in the 1660 book *Constellation Figures*. This M40 entry is actually a "negative find" for a nebulous object listed from another source. In this case, what Messier "found" at the reported location were two stars. He obtusely speculates that the author of *Constellation Figures* may have confused these stars with nebulosity. This would have been easy to do with the telescopes of the time. The indicated RA is expressed in arc degrees instead of time.

NGC Summary by J.L.E. Dreyer, circa 1888

The NGC & IC catalogues only list clusters, nebulae and galaxies.

Location

Constellation: Ursa Major (UMa)

Year 2000 Coordinates
RA: 12h 22.4m
Dec: +58° 05'

Observation Periods

Evenings 8 p.m. : February to September
Mornings 4 a.m. : October to May

Facts

Name: No common name
Type of Object: Optical double star
Magnitudes: 9.6 and 10.1
Distance: Unknown
Arc Degree Separation of the Two Stars in the Sky: 52", which is slightly less than 1'
Other: To reiterate from my notes in the left column, the two stars of M40 were not an observational misidentification of a nebula by Messier, but his best guess in reporting a "negative" find of a nebulous object noted in a contemporary catalogue.

Description of M40 using a 4-inch refractor at 48x

Appears as two equal magnitude stars next to one another. Pretty in its own right.

Locating Index: Fairly easy since it is located within 1½ degrees of the star *Megrez*.

Identifying/Observing Index: Somewhat challenging because you are looking for two faint stars. Use the picture below to guide you in. Start at *Megrez*, then move and position *70 Ursae Majoris* in the center of your eyepiece. M40 should then be visible near the perimeter of your field of view.

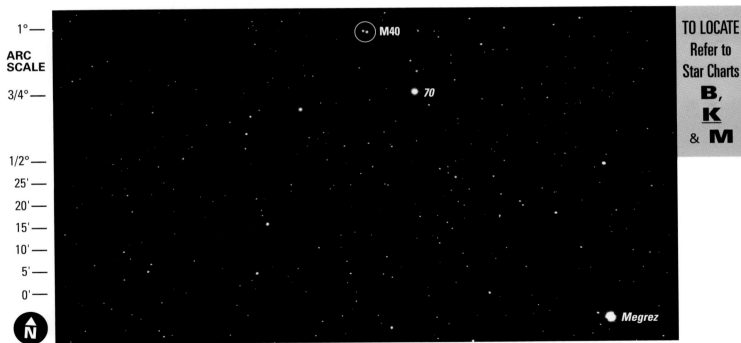

ARC SCALE

1° —
3/4° —
1/2° —
25' —
20' —
15' —
10' —
5' —
0' —

M40

70

Megrez

N

TO LOCATE
Refer to
Star Charts
**B,
K**
& **M**

The only thing that makes these two stars easy to find is their 15' distance from the 5th magnitude star *70 Ursae Majoris* which is positioned about 1½ degrees northeast of *Megrez*.

For easy comparison, all photographs are shown at the same scale, measuring 2.3° x 1.3°.

M41 Little Beehive ✧ Open Cluster

Original Messier Description

from the 1784 edition of *Connaissance des Temps*

Observed 1765. Jan. 16

Cluster of stars below *Sirius*, near o of Canis Major; this cluster appears nebulous in an ordinary one-foot telescope: it's just a cluster of small stars.

Location

Constellation: Canis Major (CMa)

Year 2000 Coordinates
RA: 6h 46.0m
Dec: −20° 44'

Observation Periods

Evenings 8 p.m. : January to April
Mornings 4 a.m. : September to December

Facts

Name: Little Beehive
Type of Object: Open cluster
Magnitude: 4½
Distance: 2,200 ly
Physical Size: Spans 24 ly
Arc Degree Size in Sky: Extends 38'
Other: Contains about 80 stars. The brightest, which shines at magnitude 6.9, is orange in color and located at its center. The age of this cluster is estimated at 190,000,000 years.

NGC Summary by J.L.E. Dreyer, circa 1888

NGC Number: 2287
Mag: 4.5   **Size:** 38'
NGC Description: Cluster, very large, bright, little compressed, contains stars of 8th magnitude and fainter.

Description of M41 using a 4-inch refractor at 48x

Spread out, stretching the length of 1½ Moons. Made of relatively bright stars, about 75 of them. Most of the stars appear to be contained within a set that roughly forms the shape of a pentagon.

Locating Index: Fairly easy because it forms the apex of a shallow isosceles triangle with *Sirius* and *o Canis Majoris*.

Identifying/Observing Index: Easy because it "pops" out at you when scanned at a lower magnification.

TO LOCATE
Refer to
Star Charts
B,
H
& **J**

ARC SCALE

1° —
3/4° —
1/2° —
25' —
20' —
15' —
10' —
5' —
0' —

N

In the picture, M41 is surrounded by lots of stars, but through a small telescope, very few of them are seen.

For easy comparison, all photographs are shown at the same scale, measuring 2.3° x 1.3°.

M42 The Great Orion Nebula ✧ Nebula

Original Messier Description

from the 1784 edition of *Connaissance des Temps*

Observed 1769. March 4

Position of the beautiful nebula in Orion's sword, from the star θ which it contains along with three other smaller stars that can only be seen with good instruments. Mr. Messier went into great detail about this big nebula; he has provided a drawing of it, done with the greatest care, which can be seen in the *Mémoires de l'Académie, year 1771, plate VIII.* It was Huyghens who discovered it in 1656. It has since been observed by very many astronomers. Reported in the English *Atlas.*

Author's Note: Huyghens is the famous Christian Huygens (1629–1695).

NGC Summary by J.L.E. Dreyer, circa 1888

NGC Number: 1976
Mag: 4 **Size:** 66'
NGC Description: A magnificent object!!! θ¹ Orionis and the great nebula.

Author's Note: *Theta (θ¹) Orionis* refers to the "star" that resolves into the Trapezium.

Location

Constellation: Orion (Ori)

Year 2000 Coordinates
RA: 5h 35.4m
Dec: −5° 27'

Observation Periods

Evenings 8 p.m. : December to April
Mornings 4 a.m. : August to December

Facts

Name: The Great Orion Nebula, Orion Nebula, Great Nebula in Orion
Type of Object: Emission nebula
Magnitude: 4
Distance: 1,500 ly
Physical Size: 29 ly x 26 ly
Arc Degree Size in Sky: 66' x 60'
Other: The gas in this nebula is being stimulated to give off its own light from energy emitted by four hot young stars, each about 1,000,000 years old, collectively called the Trapezium. The Orion Nebula bore these stars and represents a stellar nursery.

Description of M42 using a 4-inch refractor at 48x

Magnificent! Within the brightest part of the nebula are four stars, shaped somewhat like a trapezoid, thus named the Trapezium. They are well resolved at this magnification. A dark, U-shaped "bay" hugs close by. Many details within the nebula are visible, including numerous wisps.

Locating Index: Easy because it makes up part of the sword of Orion, which is visible to the naked eye.

Identifying/Observing Index: Easy because of it brightness and size. Visible in light-polluted skies.

ARC SCALE

1° —
3/4° —
1/2° —
25' —
20' —
15' —
10' —
5' —
0' —

1977

M43

The four stars that make up the Trapezium lie within the brightest part and are arranged like this:

M42

Trapezium — need 50x to see easily

TO LOCATE
Refer to
Star Charts
B,
F
& **H**

Within the sword of Orion lies M42, a large and bright nebula that is visible with the naked eye. M43 is the part of the Great Nebula that looks like a ball. Just ½ degree north looms the fainter nebula NGC 1977. For easy comparison, all photographs are shown at the same scale, measuring 2.3° x 1.3°.

M43 ✦ Nebula

Original Messier Description

from the 1784 edition of *Connaissance des Temps*

Observed 1769. March 4

Position of the little star that is surrounded by nebulosities & that is below the nebula in Orion's sword. Mr. Messier has included it in the drawing of the big one.

Location

Constellation: Orion (Ori)

Year 2000 Coordinates
RA: 5h 35.6m
Dec: −5° 16'

Observation Periods

Evenings 8 p.m. : December to April
Mornings 4 a.m. : August to December

Facts

Name: No common name but it is part of the Great Orion Nebula
Type of Object: Emission nebula
Magnitude: 9
Distance: 1,500 ly
Physical Size: 9 ly x 7 ly
Arc Degree Size in Sky: 20' x 15'
Other: M43 is being stimulated into giving off its own light by a 6.9 magnitude star.

NGC Summary   by J.L.E. Dreyer, circa 1888

NGC Number: 1982
Mag: 9   **Size:** 20'
NGC Description: Remarkable! Very bright, very large, round with tail, much brighter in the middle and containing a star of magnitude 8 or 9.

Description of M43 using a 4-inch refractor at 48x

Physically adjacent to M42. However, most observers would consider M43 to be part of M42. It appears as a circular haze with a fairly bright star in its middle. This star is a little brighter than the second brightest star in the Trapezium (see M42).

Locating Index: Easy because it is part of the bright and large M42 nebula that makes up Orion's sword.

Identifying/Observing Index: Easy because if you are looking at M42 with low power, you are also seeing M43. The hardest part of identifying M43 is distinguishing it from M42.

TO LOCATE
Refer to
Star Charts
B,
F
& **H**

ARC SCALE: 1° — 3/4° — 1/2° — 25' — 20' — 15' — 10' — 5' — 0'

M43

N

Most of us would consider M42 and M43 to be the same nebula; however, M43 is the "knot" next to the brightest part of M42.

For easy comparison, all photographs are shown at the same scale, measuring 2.3° x 1.3°.

M44 Praesepe ✧ Open Cluster

Original Messier Description

from the 1784 edition of *Connaissance des Temps*

Observed 1769. Mar. 4

Cluster of stars known by the name nebulae of Cancer. The position given is that of the star C.

Author's Note: The star "C" references a designation on a chart, a holdover from Bayer's practice of switching to Roman letters when he ran out of Greek letters. This star is perhaps one of the brighter stars in the "center" of the cluster, like *42 Cancri*. Five stars in M44 are designated with Flamsteed numbers (numbered 38 through 42).

NGC Summary by J.L.E. Dreyer, circa 1888

NGC Number: 2632
Mag: 3.1   **Size:** 95'
NGC Description: Praesepe.

Author's Note: This cluster was so well known that Dreyer's only description was its popular name.

Location

Constellation: Cancer (Cnc)
Year 2000 Coordinates
RA: 8h 40.1m
Dec: +19° 59'

Observation Periods

Evenings 8 p.m. : January to June
Mornings 4 a.m. : September to February

Facts

Name: Praesepe*, Beehive Cluster
Type of Object: Open cluster
Magnitude: 3.1
Distance: 580 ly
Physical Size: Spans 16 ly
Arc Degree Size in Sky: Extends 95' or 1.6°
Other: Contains about 50 stars with the brightest shining at magnitude 6.3. Its age is estimated at 660,000,000 years. In ancient times, the constellation Leo extended much farther east and west, and M44 was considered to be its whiskers.

*Praesepe (sometimes spelled Praesaepe) is a Latin word meaning enclosure, crib, manger, stall, a haunt or hive. Praesepe has most often been associated with a beehive, but mythology has also made it a manger, pile of hay and even King Midas — depicting his banishment, with this faint patch representing donkey ears given to him as punishment for his greed.

Description of M44 using a 4-inch refractor at 48x

Completely fills the eyepiece field of view. Not as pretty as the Pleiades (M45) because it does not form an interesting pattern and its stars are not as bright. It has 50 to 60 bright stars and appears as a fuzzy patch to the naked eyes. *Great* in binoculars.

Locating Index: **Fairly easy**, especially if you have dark skies, because you can see it as a faint patch. In more light-polluted areas, it is not quite halfway between the stars *Regulus* and *Pollux*. Scanning with binoculars makes short work of finding it.

Identifying/Observing Index: **Easy** because it is large with bright members that "pop." Use low magnification to observe it.

1°—
ARC
SCALE
3/4°—
1/2°—
25'—
20'—
15'—
10'—
5'—
0'—

42

TO LOCATE
Refer to
Star Charts
B,
H
& **J**

In modern times, M44 is known as a cluster of stars, but to the ancients, its stars were unresolvable by eye so this object was widely recognized as a nebula or cloud. *42 Cancri* is noted.

For easy comparison, all photographs are shown at the same scale, measuring 2.3° x 1.3°.

M45 The Pleiades ✧ Open Cluster

Original Messier Description

from the 1784 edition of *Connaissance des Temps*

Observed 1769. Mar. 4

Cluster of stars, known by the name *Pleiades*. The position reported is that of the star *Alcyone*.

Author's Mythological Note: The Pleiades or Seven Sisters are the daughters of Atlas and Pleione. They were changed into doves and sent into the heavens as stars to avoid the amorous clutches of Orion. Thus, the Seven Sisters always rise before Orion, forever escaping him.

One Native American legend has it that there were Seven Sisters who longed to wander among the stars, lost their way home and huddled together so as not to get separated.

The seventh star is difficult to see with the naked eye and in both stories it is said that crying by that sister blurs its brightness.

The Hyades are piglets and the half-sisters of the Pleiades, all having Atlas as their father. Together, they make up the 14 Atlantides which reside in Taurus.

NGC Summary by J.L.E. Dreyer, circa 1888

The Pleiades, like the Hyades, are open clusters but are not listed in the NGC or IC catalogues because they are large and plainly visible to the naked eye.

Location

Constellation: Taurus (Tau)

Year 2000 Coordinates
RA: 3h 47.0m
Dec: +24° 07'

Observation Periods

Evenings 8 p.m. : October to April
Mornings 4 a.m. : June to December

Facts

Name: Pleiades, the Seven Sisters
Type of Object: Open cluster
Magnitude: 1.2
Distance: 395 ly
Physical Size: Spans 13 ly
Arc Degree Size in Sky: Extends 110' or 1.8°
Other: Contains about 100 stars, with the brightest, *Alcyone,* shining at magnitude 2.9. Its age is estimated at 78,000,000 years. The Merope Nebula surrounds these stars. However, this nebula is not the cloud out of which these stars formed, but simply another one they are passing through.

Description of M45 using a 4-inch refractor at 48x

Fills the entire eyepiece field of view! Exceedingly bright stars in a pretty pattern. The nebulosity is faint and looks more like lighter patches of sky surrounding a few of the stars. Visually, this is the largest, brightest Messier object.

Locating Index: **Easy** because you can see this cluster with the naked eye. A winter sky favorite.

Identifying/Observing Index: **Easy** because it's visible to the naked eye, appearing as a little group of stars — most of us can count six of them.

ARC SCALE

1°

3/4°

1/2°
25'
20'
15'
10'
5'
0'

TO LOCATE
Refer to
Star Charts
B,
F
& **H**

Pleione
Atlas
Alcyone
Maia
Asterope
Taygeta
Celaeno
Electra
Merope

The Pleiades is a spectacular open cluster. Although it resembles a little dipper, it is not the actual Little Dipper or Ursa Minor.

For easy comparison, all photographs are shown at the same scale, measuring 2.3° x 1.3°.

M46 ✧ Open Cluster

Original Messier Description

from the 1784 edition of *Connaissance des Temps*

Observed 1771. Feb. 19

Cluster of very small stars, between the head of Canis Major & the two hind feet of Monoceros, determined by comparing this cluster with the 2nd star of Navis, 6th magnitude according to Flamsteed; these stars can only be seen with a good telescope; the cluster contains some nebulosity.

Author's Note: The constellation Navis that Messier refers to was the constellation Argo Navis or Argo, which means "The Ship of the Argonauts," captained by Jason who searched for the Golden Fleece. The constellation Argo was eventually subdivided into three constellations. Vela represents its sail, Puppis its stern and Carina its keel. From Greece, the set of stars that comprised Argo skimmed the southern horizon giving the illusion of a ship sailing on water.

NGC Summary by J.L.E. Dreyer, circa 1888

NGC Number: 2437
Mag: 6.1   **Size:** 27'
NGC Description: Remarkable! Cluster, very bright, very rich in stars, very large, involving a planetary nebula.

Location

Constellation: Puppis (Pup)
Year 2000 Coordinates
RA: 7h 41.8m
Dec: −14° 49'

Observation Periods

Evenings 8 p.m. : January to May
Mornings 4 a.m. : September to January

Facts

Name: No common name
Type of Object: Open cluster
Magnitude: 6.1
Distance: 5,400 ly
Physical Size: Spans 42 ly
Arc Degree Size in Sky: Extends for 27'
Other: Contains about 100 stars, the brightest shining at magnitude 8.7. Its age is estimated at 300,000,000 years. The planetary nebula NGC 2438 lies in front of this cluster and can be seen with averted vision in smaller telescopes.

Description of M46 using a 4-inch refractor at 48x

Almost appears as an extension of M47 because there is a trail of stars that "links" these two clusters. About half the diameter of M47 but with many more fainter stars, possibly 100 or so — with one brighter near its middle. I can just see the planetary nebula with averted vision.

Locating Index: **Slightly challenging.**

Look for the brighter M47 first by pointing the telescope at twice the distance that *γ Canis Majoris* is from *Sirius* and on the same line formed by these two stars.

Identifying/Observing Index: **Fairly easy**

in darker skies. However, it fades in light pollution. Look for the brighter M47 and then move the telescope about one degree east to find this fainter partner.

TO LOCATE
Refer to
Star Charts
B,
H
& **J**

M46 is a pretty sprinkle of stars not far from the brighter M47. The planetary nebula NGC 2438 is noted.

For easy comparison, all photographs are shown at the same scale, measuring 2.3° x 1.3°.

M47 ✦ Open Cluster

Original Messier Description

from the 1784 edition of *Connaissance des Temps*

Observed 1771. Feb. 19

Cluster of stars not far from the preceding, the stars bigger; the middle of the cluster compared to the same star, the second one of Navis. The cluster contains no nebulosity.

Author's Notes: Navis refers to the constellation Argo Navis which was broken up into the constellations Vela, Puppis and Carina. See more about this in the note for M46.

Messier's original position for M47 was incorrect because no object matched up with his coordinates. Both Oswald Thomas in 1934 and Dr. T. F. Morris in 1959 independently came to the conclusion that M46 was NGC 2422. However, Messier's description leaves little doubt as to the identity of this cluster.

NGC Summary by J.L.E. Dreyer, circa 1888

NGC Number: 2422
Mag: 4.4   **Size:** 30'
NGC Description: Cluster, bright, very large, pretty rich in stars, contains stars bright and faint.

Location

Constellation: Puppis (Pup)
Year 2000 Coordinates
RA: 7h 36.6m
Dec: −14° 30'

Observation Periods

Evenings 8 p.m. : January to May
Mornings 4 a.m. : September to January

Facts

Name: No common name
Type of Object: Open cluster
Magnitude: 4.4
Distance: 1,800 ly
Physical Size: Spans 16 ly
Arc Degree Size in Sky: Extends 30'
Other: Contains about 30 stars with the brightest shining at magnitude 5.7. Its age is estimated at 78,000,000 years.

Description of M47 using a 4-inch refractor at 48x

Three bright stars dominate, with one being a "double." Overall, it's shaped somewhat like a triangle, large and bright but sparse and "loose" with stars.

Locating Index: **Slightly challenging** because it is in an area where there are no conspicuous stars. In light-polluted skies, point the telescope at twice the distance that γ *Canis Majoris* is from *Sirius* and on the same line formed by these two stars. In dark skies, use several of the nearby 5th magnitude stars in Puppis to find it.

Identifying/Observing Index: **Fairly easy** because its bright members make it "pop" when you come across it.

TO LOCATE
Refer to
Star Charts
B,
H
& **J**

ARC SCALE

1° —
3/4° —
1/2° —
25' —
20' —
15' —
10' —
5' —
0' —

2423

M46

N

M46 and M47 are neighbors, with M47 containing brighter members, thus making it more visible. Try it using binoculars. Another cluster, NGC 2423, lies a Moon's diameter north of M47.

For easy comparison, all photographs are shown at the same scale, measuring 2.3° x 1.3°.

M48 ✧ Open Cluster

Original Messier Description

from the 1784 edition of *Connaissance des Temps*

Observed 1771. Feb. 19

Cluster of very small stars, with no nebulosity; this cluster is not far from the three stars located at the base of the tail of Monoceros.

Author's Note: Messier's original position for M48 was incorrect because no object matched up with his coordinates. Professor Owen Gingerich of Harvard and others independently concluded that M48 was probably object NGC 2548 because it has the same Right Ascension that Messier published for this object. Messier most likely mixed up the reference star when determining the cluster's Declination. A reference star is a star whose coordinates are accurately known and used as an "anchor" for determining the coordinates of other nearby stars or objects.

NGC Summary by J.L.E. Dreyer, circa 1888

NGC Number: 2548
Mag: 5.8 **Size:** 54'
NGC Description: Cluster, very large, pretty rich in stars, pretty much compressed, contains stars of magnitude 9 to 13.

Location

Constellation: Hydra (Hya)
Year 2000 Coordinates
RA: 8h 13.8m
Dec: −5° 48'

Observation Periods

Evenings 8 p.m. : January to May
Mornings 4 a.m. : September to January

Facts

Name: No common name
Type of Object: Open cluster
Magnitude: 5.8
Distance: 1,500 ly
Physical Size: Spans 24 ly
Arc Degree Size in Sky: Extends 54'
Other: Contains about 80 stars, the brightest shining at magnitude 8.2. Its age is estimated at 300,000,000 years.

Description of M48 using a 4-inch refractor at 48x

Nice cluster, large in size with fairly bright members that are spread out. It's arranged somewhat in the shape of an "A" with anywhere from 50 to 75 stars.

Locating Index: More **challenging** than most because it is not near any conspicuously bright stars. It roughly forms the apex of an isosceles triangle with *Procyon* and *Alphard* of Hydra. Scan with the telescope until you come across this bright cluster.

Identifying/Observing Index: Fairly easy because it "pops" when you come across it. It is the brightest object in this area.

TO LOCATE
Refer to
Star Charts
B,
H
& **J**

M48 is a nice large and loose cluster with bright members. The worst thing about this object is finding it, because it is in a region of the sky with mostly 4th and 5th magnitude stars.

For easy comparison, all photographs are shown at the same scale, measuring 2.3° x 1.3°.

203

M49 ✧ Virgo Cluster Galaxy

Original Messier Description

from the 1784 edition of *Connaissance des Temps*

Observed 1771. Feb. 19

Nebula discovered near the star ρ of Virgo. It's not easy to see it with an ordinary telescope of 3 & a half feet. The Comet of 1779 was compared to this nebula by Mr. Messier on April 22 & 23: The Comet & the Nebula had the same brightness. Mr. Messier reported this nebula on the Chart of the track of this Comet, which will appear in the volume of the Academy for the same year, 1779. Observed again on April 10, 1781.

NGC Summary by J.L.E. Dreyer, circa 1888

NGC Number: 4472
Mag: 8.4 **Size:** 8.9'
NGC Description: Very bright, large, round, much brighter in the middle, not resolvable.

Location

Constellation: Virgo (Vir)

Year 2000 Coordinates
RA: 12h 29.8m
Dec: +8° 00'

Observation Periods

Evenings 8 p.m. : March to August
Mornings 4 a.m. : November to April

Facts

Name: No common name
Type of Object: Elliptical galaxy
Magnitude: 8.4
Distance: 56,000,000 ly
Physical Size: Spans at least 147,000 ly
Arc Degree Size in Sky: 9' x 7'
Other: Classified as an E4 elliptical, which means that it has a medium elongation. This galaxy is a member of the Virgo Galaxy Cluster.

Description of M49 using a 4-inch refractor at 48x

Bright and fairly large galaxy whose elliptical nature is easy to recognize. Its core looks like a star! Located between two 6th magnitude stars that are spaced almost 1½° apart, with M49 closer to one than the other.

Locating Index: Challenging because it is not near any conspicuously bright stars. To locate this galaxy, use Chart U, starting at the star *Vindemiatrix* and then "star hop" until you find it.

Identifying/Observing Index: Fairly easy once you locate its position because it is bright for a galaxy. You will be able to see it at many light-polluted locations.

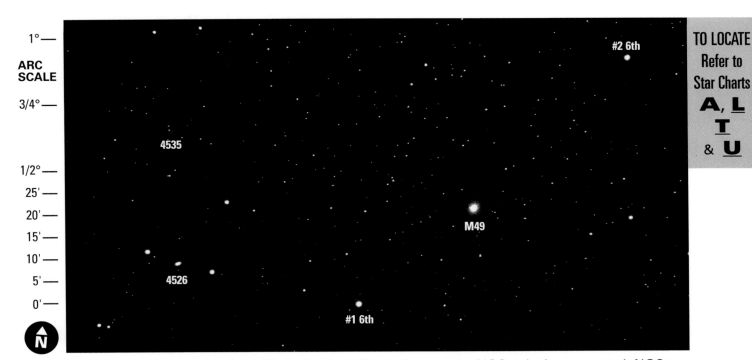

ARC SCALE

1° —
3/4° —
1/2° —
25' —
20' —
15' —
10' —
5' —
0' —

4535
4526
M49
#2 6th
#1 6th

N

TO LOCATE
Refer to
Star Charts
A, L
T
& **U**

M49 is a good example of an elliptical galaxy. Two other nearby NGC galaxies are noted. NGC 4526 is very bright and centered between two bright stars. Did Messier see this as just a star?

For easy comparison, all photographs are shown at the same scale, measuring 2.3° x 1.3°.

M50 ✧ Open Cluster

Original Messier Description

from the 1784 edition of *Connaissance des Temps*

Observed 1772. Apr. 5

Cluster of stars, more or less brilliant, below the right thigh of Monoceros, above the star θ of the ear of Canis Major, & near a 7th magnitude star. It was while observing the Comet of 1772 that Mr. Messier observed this cluster. He reported it on the Chart of that Comet, which he traced. *Mém. Acad. 1772.*

Location

Constellation: Monoceros (Mon)

Year 2000 Coordinates
RA: 7h 02.8m
Dec: −8° 23'

Observation Periods

Evenings 8 p.m. : January to May
Mornings 4 a.m. : September to January

Facts

Name: No common name
Type of Object: Open cluster
Magnitude: 5.9
Distance: 3,000 ly
Physical Size: Spans 14 ly
Arc Degree Size in Sky: Extends 16'
Other: Contains about 80 stars, the brightest shining at magnitude 7.9. Its age is estimated at 78,000,000 years.

NGC Summary by J.L.E. Dreyer, circa 1888

NGC Number: 2323
Mag: 5.9 **Size:** 16'
NGC Description: Remarkable! Cluster, very large, rich in stars, pretty compressed, extended, contains stars of magnitude 12 to 16.

Description of M50 using a 4-inch refractor at 48x

In an area of the Milky Way Band thick with stars, This cluster does not stand out like others. Contains one star that is brighter than the rest. Around 20' in diameter. Can see more stars with averted vision. It's shaped like a spiral galaxy with arms radiating from the center.

Locating Index: Slightly challenging because it is not near any conspicuously bright stars. However, it is on a line between the stars *Sirius* and *Procyon*.

Identifying/Observing Index: Slightly challenging because it does not "pop" out at you like other clusters. Although it is the brightest cluster in the area, you may sweep past it a few times before you realize its identity.

ARC
SCALE

1° —
3/4° —
1/2° —
25' —
20' —
15' —
10' —
5' —
0' —

TO LOCATE
Refer to
Star Charts
B,
H
& **J**

M50 is in a part of the Milky Way Band that has a good concentration of stars. Don't be surprised if you pass this cluster by a few times before identifying it.

For easy comparison, all photographs are shown at the same scale, measuring 2.3° x 1.3°.

M 51 Whirpool Galaxy ✧ Galaxy

Original Messier Description

from the 1784 edition of *Connaissance des Temps*

Observed 1774. Jan. 11

Very faint nebula, without stars, near the ear of Canes Venatici, the northernmost one, below the star η, 2nd magnitude, of Ursa Major's tail. Mr. Messier discovered this nebula on October 13, 1773, while observing the Comet that appeared that year. It can only be seen with great difficulty with an ordinary 3½ foot telescope; near it, there is an 8th magnitude star. Mr. Messier reported its position on the Chart of the Comet observed in 1773 & 1774. Mémoires de l'Académie 1774, plate III. It's double, each has a bright center, and these are 4' 35" away from each other. The two atmospheres are contiguous, one is fainter than the other. Observed again several times.

NGC Summary by J.L.E. Dreyer, circa 1888

NGC Number: 5194
Mag: 8.4   **Size:** 11'
NGC Description: A magnificent object!!! The great spiral nebula M51.

Location

Constellation: Canes Venatici (CVn)
Year 2000 Coordinates
RA: 13h 29.9m
Dec: +47° 12'

Observation Periods

Evenings 8 p.m. : March to September
Mornings 4 a.m. : November to May

Facts

Name: Whirlpool Galaxy, Rosse's Galaxy and (Rosse's) Question Mark Galaxy*
Type of Object: Spiral galaxy
Magnitude: 8.1
Distance: 37,000,000 ly
Physical Size: 118,000 ly in diameter
Arc Degree Size in Sky: 11' x 8'
Other: The shape of this galaxy is classified as Sc, which means that it has a round nucleus with loose open arms. The bright "Knot" off its arm is the more distant galaxy NGC 5195 that had once interacted with the Whirlpool Galaxy.

*Rosse is Lord Rosse of Birr Castle in Ireland — see page 52.

Description of M51 using a 4-inch refractor at 48x

In "average" skies, it looks like two stars with some nebulosity surrounding each. One star is the "Knot" and the other star is in the center of the Whirlpool. Under darker skies, the arms are faint but clearly visible.

Locating Index: **Fairly easy** because it roughly forms the base of a right triangle with the stars *Alkaid* and *Mizar*.

Identifying/Observing Index: **Easy** in dark skies. **Slightly challenging** in fair skies. **Difficult** in light-polluted skies.

ARC SCALE

1° —
3/4° —
1/2° —
25' —
20' —
15' —
10' —
5' —
0' —

Knot

N

TO LOCATE
Refer to
Star Charts
A,
K
& **M**

The Whirlpool is a magnificent galaxy, even with a small telescope under dark skies. It is unfortunate that Messier and others of his time did not know the true nature of this object.

For easy comparison, all photographs are shown at the same scale, measuring 2.3° x 1.3°.

M 52 The Scorpion ✧ Open Cluster

Original Messier Description

from the 1784 edition of *Connaissance des Temps*

Observed 1774. Sept. 7

Cluster of very small stars mingled with nebulosity which can only be seen with an achromatic telescope. It was while observing the Comet which appeared that year that Mr. Messier saw this cluster which was close to the Comet on September 7th, 1774. It's below the star d of Cassiopeia; this star d was used to determine the cluster of stars & the Comet.

NGC Summary by J.L.E. Dreyer, circa 1888

NGC Number: 7654
Mag: 6.9   **Size:** 13'
NGC Description: Cluster, large, rich in stars, much compressed in the middle, round, contains stars of magnitude 9 to 13.

Location

Constellation: Cassiopeia (Cas)

Year 2000 Coordinates
RA: 23h 24.2m
Dec: +61° 35'

Observation Periods

Evenings 8 p.m. : August to February
Mornings 4 a.m. : April to October

Facts

Name: The Scorpion
Type of Object: Open cluster
Magnitude: 6.9
Distance: 3,000 ly
Physical Size: Spans 11 ly
Arc Degree Size in Sky: Extends 13'
Other: Contains at least 100 stars, the brightest shining at magnitude 8.2. Its age is estimated at 35,000,000 years.

Description of M52 using a 4-inch refractor at 48x

Straddling two relatively bright stars, it bears some resemblance to M103, but it is slightly smaller. With averted vision, it appears triangular in shape.

Locating Index: Fairly easy because it is on the same line formed by *Caph* and *α Cassiopeiae* and it is the same distance from *Caph* as *Caph* is from *α Cassiopeiae*.

Identifying/Observing Index: Fairly easy because it "pops" into view compared to M103, for the area immediately surrounding it is sparse with stars.

TO LOCATE
Refer to
Star Charts
**A,
C**
& **Q**

The looser cluster of stars below M52 is known as Czernik 43 (see more about designations in the sidebar on page 54).

For easy comparison, all photographs are shown at the same scale, measuring 2.3° x 1.3°.

M53 ✧ Globular Cluster

Original Messier Description

from the 1784 edition of *Connaissance des Temps*

Observed 1777. Feb. 26

Nebula without stars, discovered below & near Coma Berenices, not far from the 42nd star of that constellation, according to Flamsteed. This nebula is round & visible. The Comet of 1779 was directly compared to this nebula, & Mr. Messier has reported it on the Chart of that Comet, which will be inserted in the volume of the Académie of 1779. Observed again on April 13, 1781: It resembles the nebula which is below Lepus.

NGC Summary by J.L.E. Dreyer, circa 1888

NGC Number: 5024
Mag: 7.7   **Size:** 12.6'
NGC Description: Remarkable! Globular cluster of stars, bright, very compressed, irregularly round, *very* much becomes brighter in the middle, contains stars of magnitude 12.

Location

Constellation: Coma Berenices (Com)
Year 2000 Coordinates
RA: 13h 12.9m
Dec: +18° 10'

Observation Periods

Evenings 8 p.m. : March to August
Mornings 4 a.m. : November to April

Facts

Name: No common name
Type of Object: Globular cluster
Magnitude: 7.6
Distance: 60,000 ly
Physical Size: 220 ly in diameter
Arc Degree Size in Sky: 12.6' in diameter
Other: This globular cluster has a slightly tighter compactness than average.

Description of M53 using a 4-inch refractor at 48x

A slightly smaller and slightly fainter version than the wonderful M3, otherwise these two globulars are similar in appearance. It has a bright, extended core.

Locating Index: Slightly challenging because it is in an area with no conspicuously bright stars. Try finding the 4th magnitude, bottom, *a* star of Coma Berenices because at low magnification, it is only an eyepiece field of view away.

Identifying/Observing Index: Easy because it is bright, so it "pops" when you come across it.

M53

TO LOCATE
Refer to
Star Charts
A,
L
& **T**

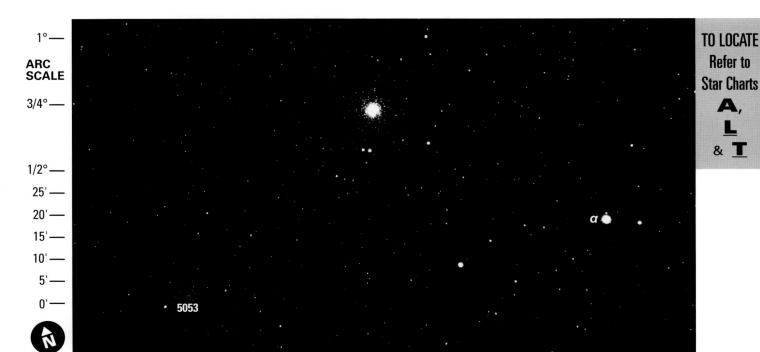

M53 is 1 degree northeast of *a Coma Berenices*. A smaller and fainter globular cluster, NGC 5053 is noted (positioned just above the number).

For easy comparison, all photographs are shown at the same scale, measuring 2.3° x 1.3°.

213

M54 ✦ Globular Cluster

Original Messier Description
from the 1784 edition of *Connaissance des Temps*

Observed 1778. July 24

Very faint nebula, discovered in Sagittarius. Its center is bright & contains no stars, seen with an achromatic telescope of 3½ feet. Its position was determined from ζ of Sagittarius, 3rd magnitude.

NGC Summary   by J.L.E. Dreyer, circa 1888
NGC Number: 6715
Mag: 7.7   **Size:** 9.1'
NGC Description: Globular cluster of stars, very bright, large, round, gradually then suddenly becomes much brighter in the middle, well resolved, contains stars of magnitude 15.

Location
Constellation: Sagittarius (Sgr)
Year 2000 Coordinates
RA: 18h 55.1m
Dec: –30° 29'

Observation Periods
Evenings 8 p.m. : August to October
Mornings 4 a.m. : April to June

Facts
Name: No common name
Type of Object: Globular cluster
Magnitude: 7.6
Distance: Estimates range from 49,000 ly to 89,000 ly
Physical Size: 128 ly in diameter for the closest distance estimate, and 233 ly for the farthest
Arc Degree Size in Sky: 9' in diameter
Other: The stars in this globular cluster are compacted very tightly. If the data on this globular is correct (especially the farther distance estimate), then this globular would be the largest in size attached to our galaxy.

Description of M54 using a 4-inch refractor at 48x

Pretty compact, small in comparison to M22. Its center is starlike and fairly bright. Similar to M14. This is the brightest of the three globulars beneath Sagittarius.

Locating Index: Fairly easy to find because it it almost on a line between ζ *Sagittarii* and *Kaus Australis* and less than ¼ of this distance from ζ *Sagittarii*.

Identifying/Observing Index: Fairly easy because its center is obvious since it is starlike and bright. **Slightly challenging** in light-polluted skies.

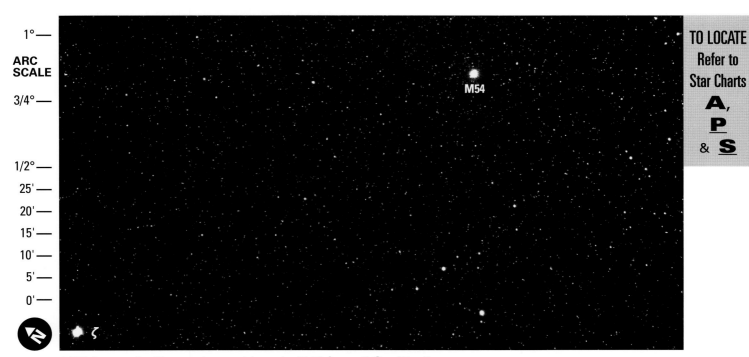

TO LOCATE
Refer to
Star Charts
A,
P
& **S**

1° —
ARC SCALE
3/4° —
1/2° —
25' —
20' —
15' —
10' —
5' —
0' —

M54

ζ

M54 has a starlike center and is only 1½° from ζ *Sagittarii*.

For easy comparison, all photographs are shown at the same scale, measuring 2.3° x 1.3°.

M55 The Spectre ✦ Globular Cluster

Original Messier Description
from the 1784 edition of *Connaissance des Temps*

Observed 1778. July 24

Nebula which is a whitish spot, with an expanse of about 6 minutes, its light is even & didn't appear to contain any stars. Its position was determined from ζ of Sagittarius, by means of a 7th magnitude star between them. This nebula was discovered by Father de la Caille. *Mém. Acad. 1755, p. 194.* Mr. Messier looked for it in vain on July 29, 1764, as he reports in his Mémoire.

Location
Constellation: Sagittarius (Sgr)
Year 2000 Coordinates
RA: 19h 40.0m
Dec: –30° 58'

Observation Periods
Evenings 8 p.m. : August to October
Mornings 4 a.m. : April to June

Facts
Name: The Spectre* (of M22)
Type of Object: Globular cluster
Magnitude: 7.0
Distance: 20,000 ly
Physical Size: 111 ly in diameter
Arc Degree Size in Sky: 19' in diameter
Other: The compactness of the stars in M55 is extremely loose in comparison to other globular clusters.

*Spectre is a variation on the spelling of the word "specter." Spectre is also the French spelling of specter, and means wraith or ghost.

Description of M55 using a 4-inch refractor at 48x

Great globular that is big and similar to M22 but with half the brightness. I can see individual stars. Its luminosity is fairly even across its diameter; however, its center is brighter and extended. One of my favorites.

Locating Index: Slightly challenging because it is not near any conspicuously bright stars. It is on the same line formed by τ *Sagittarii* and *Nunki*, but twice as far from τ *Sagittarii* as τ *Sagittarii* is from *Nunki*.

Identifying/Observing Index: Easy because it "pops" when you come across it, but it will be more difficult to see in light-polluted skies.

ARC SCALE
1° —
3/4° —
1/2° —
25' —
20' —
15' —
10' —
5' —
0' —

TO LOCATE
Refer to Star Charts
A, P & R

M55 is a counterpart to M22 but darker and more mysterious looking.

For easy comparison, all photographs are shown at the same scale, measuring 2.3° x 1.3°.

M56 ✦ Globular Cluster

Original Messier Description

from the 1784 edition of *Connaissance des Temps*

Observed 1779. Jan. 23

Nebula without stars, with little light;
Mr. Messier discovered it on the very
same day that he found the Comet of
1779, January 19. On the 23rd, he determined its position by comparing it to the
star No. 2 of Cygnus, according to Flamsteed: it's near the Milky Way; there is
a tenth magnitude star close to it. Mr.
Messier reported it on the Chart of the
Comet of 1779.

NGC Summary by J.L.E. Dreyer, circa 1888

NGC Number: 6779
Mag: 8.3 **Size:** 7.1'
NGC Description: Globular cluster of
stars, bright, large, irregularly round,
gradually becomes very much compressed in the middle, well resolved,
contains stars of magnitude 11 to 14.

Location

Constellation: Lyra (Lyr)
Year 2000 Coordinates
RA: 19h 16.6m
Dec: +30° 11'

Observation Periods

Evenings 8 p.m. : June to December
Mornings 4 a.m. : February to August

Facts

Name: No common name
Type of Object: Globular cluster
Magnitude: 8.3
Distance: 33,000 ly
Physical Size: 67 ly in diameter
Arc Degree Size in Sky: 7' in diameter
Other: The compactness of the stars
in M56 is very loose in comparison
to other globular clusters, and just
slightly tighter than M55.

Description of M56 using a 4-inch refractor at 48x

A fainter globular with a fairly bright center that fades outward.

Locating Index: **Fairly easy** because M56 is about halfway between *Albireo* and *γ Lyrae* or *Sulafat*.

Identifying/Observing Index: **Slightly challenging** because it is not very bright. It also gets lost in this thick part of the Milky Way. You will have problems seeing this one in light-polluted skies.

TO LOCATE
Refer to
Star Charts
A,
Q
& **R**

M56, located at the center of this picture, does not particularly stand out because it is not very bright. Additionally, it often gets sidestepped because it is in an area of the sky filled with spectacular Messier objects.

For easy comparison, all photographs are shown at the same scale, measuring 2.3° x 1.3°.

M57 Ring Nebula ✧ Planetary Nebula

Original Messier Description

from the 1784 edition of *Connaissance des Temps*

Observed 1779. Jan. 31

Mass of light located between γ & β of Lyra, discovered while looking for the Comet of 1779 which passed very close to it. It seemed like this mass of light, which is rounded, was composed of very small stars; with the best telescope it's impossible to see them, one merely suspects that there are some. Mr. Messier reported this mass of light on the Chart of the Comet of 1779. Mr. Darquier, in Toulouse, discovered this nebula while observing the same Comet, & he reports: "Nebula between γ & β of Lyra; it's very dull, but perfectly outlined; it's as large as Jupiter & looks like a dying Planet."*

NGC Summary by J.L.E. Dreyer, circa 1888

NGC Number: 6720
Mag: 9.0 **Size:** 2.5'
NGC Description: A magnificent object!!! Ring nebula, bright, pretty large, considerably extended (in Lyra).

Location

Constellation: Lyra (Lyr)
Year 2000 Coordinates
RA: 18h 53.6m
Dec: +33° 02'

Observation Periods

Evenings 8 p.m. : June to December
Mornings 4 a.m. : February to August

Facts

Name: Ring Nebula
Type of Object: Planetary nebula
Magnitude: 9
Distance: Estimates range from 1,400 ly to 5,000 ly
Physical Size: About ½ ly in length at the closest distance estimate and 1.9 ly for the farthest
Arc Degree Size in Sky: 1.3' in length
Other: Since the Hubble Space Telescope took spectacular pictures of M57, I have read that the actual shape of this planetary is a cylinder, a shell or a ring. Take your pick.

*Author's Note: Antoine Darquier de Pellepoix (1718–1802) was a French astronomer who collaborated with Lalande on a star catalogue.

Description of M57 using a 4-inch refractor at 48x

Small, well defined, doughnutlike nebula. One of the few deep sky objects that appears to have a definite edge. It is evenly illuminated and glimpses of the central, dark hole can be seen. It may not seem like it, but M57 is bigger than Jupiter or Venus at their largest. A favorite object.

Locating Index: Easy because it is positioned between the two end stars of Lyra's parallelogram — a little closer than halfway to β.

Identifying/Observing Index: Easy or **challenging**. Easy to spot if you have observed it before, but it could prove difficult if you are trying to find it for the first time, because it may be smaller and fainter than you expect. It is visible in many light-polluted skies.

TO LOCATE
Refer to Star Charts
A, **O** & **P**

The Ring Nebula, centered, is a unique deep sky object because it has a "sharp" edge. Although it is measured as one of the smallest Messier objects, it appears larger than many of the Messier galaxies because it is evenly illuminated.

For easy comparison, all photographs are shown at the same scale, measuring 2.3° x 1.3°.

M58 ✦ Virgo Cluster Galaxy

Original Messier Description

from the 1784 edition of *Connaissance des Temps*

Observed 1779. April 15

Very faint nebula discovered in Virgo, almost on the parallel of ε, 3rd magnitude. The slightest light to illuminate the micrometer wires made it disappear. Mr. Messier reported it on the Chart of the Comet of 1779, which will be in the Academy volume for the same year.

Author's Note: The micrometer wires Messier refers to were "reticle lines" visible through the eyepiece. These wires would have been used to increase the accuracy of positional measurement. Since batteries were not invented until about 1800, it is most likely that he used the light from an oil lamp to illuminate the wires (maybe through a hole in the side of the eyepiece) and turned the wick "up" or "down" to change brightness.

NGC Summary by J.L.E. Dreyer, circa 1888

NGC Number: 4579
Mag: 9.8   **Size:** 5.4'
NGC Description: Bright, large, irregularly round, very much brighter in the middle, not resolved.

Location

Constellation: Virgo (Vir)

Year 2000 Coordinates
RA: 12h 37.7m
Dec: +11° 49'

Observation Periods

Evenings 8 p.m. : March to August
Mornings 4 a.m. : November to April

Facts

Name: No common name
Type of Object: Spiral galaxy
Magnitude: 9.8
Distance: 56,000,000 ly
Physical Size: 81,000 ly in diameter
Arc Degree Size in Sky: 5' x 4'
Other: The shape of this spiral is classified as Sb, which means that it has a round nucleus with medium wound arms. It's a member of the Virgo Galaxy Cluster.

Description of M58 using a 4-inch refractor at 48x

Of the M58, M59 and M60 trio that spans 1½ degrees, M58 is similar in size and brightness to M59, but both are half the size and brightness of M60. M58 appears somewhat elongated, meaning that its orientation is tilted (from a face-on orientation) to our view.

Locating Index: Challenging because this object is small and faint. Additionally, it is not near any conspicuously bright stars. Start at the star *Vindemiatrix*, then using Chart U, star hop until you locate it. Use the brighter M60 as an area anchor.

Identifying/Observing Index: Fairly easy in dark skies but it will prove **challenging** in skies with light pollution.

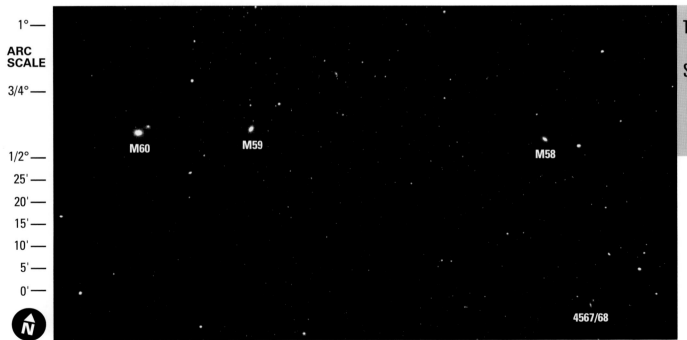

TO LOCATE Refer to Star Charts **A, L, T** & **U**

Messier referenced these three galaxies together because they are on the "same" Declination as the star *Vindemiatrix*. The galaxies NGC 4567 & 4568 are commonly known as the Siamese Twins.

For easy comparison, all photographs are shown at the same scale, measuring 2.3° x 1.3°.

M 59 ✧ Virgo Cluster Galaxy

Original Messier Description

from the 1784 edition of *Connaissance des Temps*

Observed 1779. April 15

Nebula in Virgo & in the vicinity of the preceding one, on the parallel of ε, which was used to determine it. It has the same brightness as the one mentioned above, just as faint. Mr. Messier reported it on the Chart of the Comet of 1779.

Location

Constellation: Virgo (Vir)

Year 2000 Coordinates
RA: 12h 42.0m
Dec: +11° 39'

Observation Periods

Evenings 8 p.m. : March to August
Mornings 4 a.m. : November to April

Facts

Name: No common name
Type of Object: Elliptical galaxy
Magnitude: 9.8
Distance: 56,000,000 ly
Physical Size: Spans at least 81,000 ly
Arc Degree Size in Sky: 5' x 3'
Other: The shape of this galaxy is classified as E3, which means that it is less than "average" in elongation (see page 340). It is a member of the Virgo Galaxy Cluster.

NGC Summary by J.L.E. Dreyer, circa 1888

NGC Number: 4621
Mag: 9.8 **Size:** 5.1'
NGC Description: Bright, pretty large, little elongated, very suddenly becomes very much brighter in the middle, 2 stars preceding westward.

Description of M59 using a 4-inch refractor at 48x

Of the M58, M59 and M60 trio that spans 1½ degrees, M59 is similar in size and brightness to M58, but both are half the size and brightness of M60. A slight elongation is apparent.

Locating Index: Challenging because this object is small and faint. Additionally, it is not near any conspicuously bright stars. Start at the star *Vindemiatrix*, then using Chart U, star hop until you locate it. Find the brighter M60 and then look 15' off its west side for M59.

Identifying/Observing Index: Fairly easy in dark skies, but it will prove **challenging** in skies with light pollution.

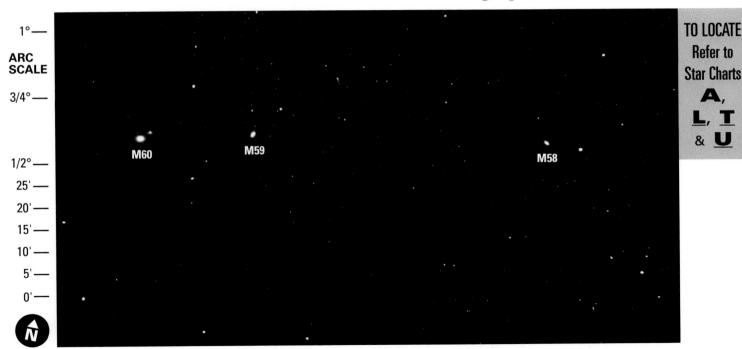

TO LOCATE
Refer to Star Charts
A, L, T & **U**

M59 is a little easier to identify than M58 because it is only 15' from the brighter M60. If you center M58 in your eyepiece with telescope-tracking off, M59 will drift onto the north/south centerline 4½ minutes later, and M60, 2 minutes after that. For easy comparison, all photographs are shown at the same scale, measuring 2.3° x 1.3°.

M60 ✦ Virgo Cluster Galaxy

Original Messier Description

from the 1784 edition of *Connaissance des Temps*

Observed 1779. April 15

Nebula in Virgo, a little more visible than the two preceding ones, also on the parallel of ε, which was used to determine it. Mr. Messier reported it on the Chart of the Comet of 1779. He discovered these three nebulae while observing the Comet, which passed very close to them. The latter [Comet] passed so close to them on April 13 & 14, that since they were both in the telescope's field, he couldn't see it. It wasn't until the 15th, while looking for the Comet, that he spotted this nebula. These three nebulae seemed to contain no stars.*

NGC Summary by J.L.E. Dreyer, circa 1888

NGC Number: 4649
Mag: 8.8   **Size:** 7.2'
NGC Description: Very bright, pretty large, round, the eastward half of a double nebula.

Author's Note: During Dreyer's time, galaxies were referred to as nebulae because their true nature was still uncertain.

Location

Constellation: Virgo (Vir)

Year 2000 Coordinates
RA: 12h 43.7m
Dec: +11° 33'

Observation Periods

Evenings 8 p.m. : March to August
Mornings 4 a.m. : November to April

Facts

Name: No common name
Type of Object: Elliptical galaxy
Magnitude: 8.8
Distance: 56,000,000 ly
Physical Size: Spans at least 114,000 ly
Arc Degree Size in Sky: 7' x 6'
Other: The shape of this galaxy is classified as E1, which means that it is close to a sphere (see page 340). It's a member of the Virgo Galaxy Cluster.

*Author's Note: The word in brackets was added by the translator for clarity.

Description of M60 using a 4-inch refractor at 60x

Very similar to M49 a little farther south. Appears starlike with some nebulosity around it. Fairly round. This is one of the brighter galaxies in this area, so use it as an anchor and for comparison.

Locating Index: Challenging because it is not near any conspicuously bright stars. Start at the star *Vindemiatrix*, then using Chart U, star hop until you locate it.

Identifying/Observing Index: Easy in dark skies, but a little more **challenging** in light-polluted skies where it will appear starlike.

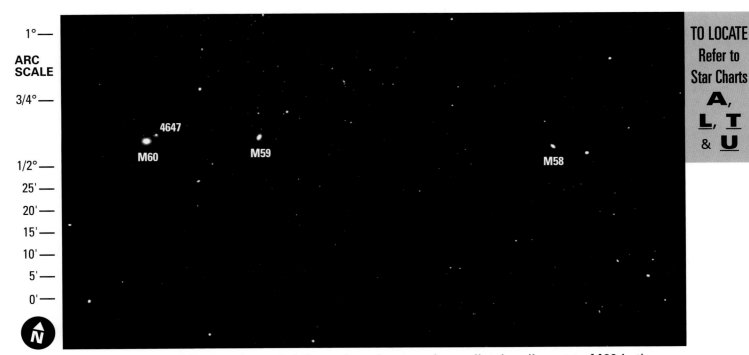

TO LOCATE Refer to Star Charts **A, L, T** & **U**

Of the trio above, M60 is twice as bright as the other two. Immediately adjacent to M60 is the faint galaxy NGC 4647. To Messier, this probably looked like a star.

For easy comparison, all photographs are shown at the same scale, measuring 2.3° x 1.3°.

M61 Swelling Spiral ✦ Virgo Cluster Galaxy

Original Messier Description

from the 1784 edition of *Connaissance des Temps*

Observed 1779. May 11

Nebula, very faint & hard to see. Mr. Messier mistook this nebula for the Comet of 1779 on the 5th, 6th, & 11th of May. On the 11th, he realized that it wasn't the Comet, but a nebula which was on its path & in the same place in the sky.

Location

Constellation: Virgo (Vir)

Year 2000 Coordinates
RA: 12h 21.9m
Dec: +4° 28'

Observation Periods

Evenings 8 p.m. : March to August
Mornings 4 a.m. : November to April

NGC Summary by J.L.E. Dreyer, circa 1888

NGC Number: 4303
Mag: 9.7 **Size:** 6.0'
NGC Description: Very bright, very large, very suddenly becomes brighter in the middle like a single star, binuclear.

Facts

Name: Swelling Spiral
Type of Object: Spiral galaxy
Magnitude: 9.7
Distance: 56,000,000 ly
Physical Size: 98,000 ly in diameter
Arc Degree Size in Sky: 6' x 5'
Other: The shape of this galaxy is classified as Sc, which means that it has a round nucleus with loosely wound arms (see page 340). It is a member of the Virgo Galaxy Cluster.

Description of M61 using a 4-inch refractor at 48x

Overall, this face-on spiral appears uniformly diffused — almost like a nebula. However, its core or nucleus does appear as a faint star. About half the size of M64, and around ½ to ⅓ of its brightness.

Locating Index: Challenging because it is not near any conspicuously bright stars. It does, however, form the right angle of a triangle with the stars *Vindemiatrix* and *Denebola*.

Identifying/Observing Index: In dark skies, **fairly easy**, but in light-polluted skies, **challenging**, so look for its starlike nucleus.

TO LOCATE
Refer to
Star Charts
**A,
L,
& T**

M61 is centered at the top of this picture. This spiral galaxy, at the "bottom" of the Virgo Galaxy Cluster pack, is just a little over 1 degree north of the 5th magnitude star *16 Virginis*.

For easy comparison, all photographs are shown at the same scale, measuring 2.3° x 1.3°.

M62 Flickering Globular ✦ Globular Cluster

Original Messier Description

from the 1784 edition of *Connaissance des Temps*

Observed 1779. Jun. 4

A very beautiful nebula, discovered in Scorpius; it resembles a little Comet. Its center is bright & surrounded by a faint glow. Its position determined by comparing it to the star τ of Scorpius. Mr. Messier had already seen this nebula on June 7, 1771, but had only determined its approximate position. Observed again on March 22, 1781.

NGC Summary by J.L.E. Dreyer, circa 1888

NGC Number: 6266
Mag: 6.6   **Size:** 14.1'
NGC Description: Remarkable! Globular cluster of stars, very bright, large, gradually becomes much brighter in the middle, well resolved, contains stars of magnitude 14 to 16.

Location

Constellation: Ophiuchus (Oph)
Year 2000 Coordinates
RA: 17h 01.2m
Dec: –30° 07'

Observation Periods

Evenings 8 p.m. : July to September
Mornings 4 a.m. : March to May

Facts

Name: Flickering Globular
Type of Object: Globular cluster
Magnitude: 6½
Distance: 22,000 ly
Physical Size: 90 ly in diameter
Arc Degree Size in Sky: 14' in diameter
Other: This globular cluster has a tighter compactness than average. It was given the name "Flickering" by Stephen O'Meara when he observed the core apparently changing in brightness and color, "like a dying flame." He attributed this effect to an optical illusion created by the arrangement of its stars.

Description of M62 using a 4-inch refractor at 48x

Pretty bright and nice globular, with a compact and bright center. Overall, it's about ⅓ the size of M4.

Locating Index: Slightly challenging. Approximately forms the right angle of a triangle with the stars ε and τ *Scorpii*.

Identifying/Observing Index: Fairly easy because it is bright and thus "pops" when you come across it.

TO LOCATE
Refer to
Star Charts
A,
P
& **S**

Messier discovered M62 after he had already catalogued numerous objects in this area of the sky.

For easy comparison, all photographs are shown at the same scale, measuring 2.3° x 1.3°.

M 63 Sunflower Galaxy ✧ Galaxy

Original Messier Description

from the 1784 edition of *Connaissance des Temps*

Observed 1779. Jun. 14

Nebula discovered by Mr. Méchain in Canes Venatici. Mr. Messier searched for it; it's faint, its brightness is nearly the same as that of the nebula reported in *No. 59*. It contains no stars, & the slightest light to illuminate the micrometer wires made it disappear. Close to it there's an 8th magnitude star, which precedes the nebula in right ascension. Mr. Messier reported its position on the Chart of the track of the Comet of 1779.

Author's Note: M63 represents Méchain's first deep sky object discovery and marks his entrance into compiling Messier's catalogue.

NGC Summary by J.L.E. Dreyer, circa 1888

NGC Number: 5055
Mag: 8.6   **Size:** 12.3'
NGC Description: Very bright, large, pretty much extended approximately along position angle 120°, very suddenly becomes much brighter in the middle to a bright nucleus.

Location

Constellation: Canes Venatici (CVn)
Year 2000 Coordinates
RA: 13h 15.8m
Dec: +42° 02'

Observation Periods

Evenings 8 p.m. : March to September
Mornings 4 a.m. : November to May

Facts

Name: Sunflower Galaxy
Type of Object: Spiral galaxy
Magnitude: 8.6
Distance: 30,000,000 ly
Physical Size: 105,000 ly in diameter
Arc Degree Size in Sky: 12' x 8'
Other: Shape classified as Sb, which means that it has a round nucleus with medium wound arms (see page 340).

Description of M63 using a 4-inch refractor at 48x

A fairly bright spiral galaxy with a star from our Milky Way Galaxy sitting on its edge. Its nucleus is almost starlike, surrounded by "nebulosity." The plane of this galaxy is angled toward us.

Locating Index: Challenging because it is not near any conspicuously bright stars, but just off a line between the stars *Alkaid* and *a Canum Venaticorum*.

Identifying/Observing Index: Fairly easy to spot in dark skies, but it will prove more **challenging** in skies with light pollution. The 9th magnitude star sitting on its edge can be used for positive identification.

TO LOCATE
Refer to
Star Charts
**A,
K**
& **M**

The Sunflower has a 9th magnitude star from our galaxy "sitting" on its edge.

For easy comparison, all photographs are shown at the same scale, measuring 2.3° x 1.3°.

M64 Black Eye Galaxy ✧ Galaxy

Original Messier Description

from the 1784 edition of *Connaissance des Temps*

Observed 1780. March 1

Nebula discovered in Coma Berenices which is about half as visible as the one located below the hair. Mr. Messier has reported its position on the Chart of the Comet of 1779. Observed again on March 17, 1781.

Location

Constellation: Coma Berenices (Com)

Year 2000 Coordinates

RA: 12h 56.7m

Dec: +21° 41'

Observation Periods

Evenings 8 p.m. : March to August

Mornings 4 a.m. : November to April

Facts

Name: Black Eye Galaxy, Evil Eye Galaxy

Type of Object: Spiral galaxy

Magnitude: 8.5

Distance: Estimates range from 12,000,000 ly to 22,800,000 ly

Physical Size: 31,000 ly in diameter for the closest distance estimate and 60,000 ly for the farthest

Arc Degree Size in Sky: 9' x 5'

Other: Shape classified as Sb, which means that it has a round nucleus with medium wound arms. A dark dust lane gives the illusion that it has a black eye.

NGC Summary by J.L.E. Dreyer, circa 1888

NGC Number: 4826

Mag: 8.5   **Size:** 9.3'

NGC Description: Remarkable! Very bright, very large, very much extended approximately along position angle 120°, bright in the middle with a small bright nucleus.

Description of M64 using a 4-inch refractor at 48x

Similar to M63. It appears as a relatively bright center, almost starlike, with lots of surrounding "nebulosity." I think I can discern two distinct bright areas near the nucleus. Overall, this galaxy is more face on and not as bright as you might expect.

Locating Index: Challenging because it is not near any conspicuously bright stars. It does form the apex of a shallow isosceles triangle with the stars *Arcturus* and *Denebola*.

Identifying/Observing Index: Easy in dark skies because it is fairly bright and "pops" but it fades in light-polluted skies where it can prove **challenging**.

TO LOCATE
Refer to
Star Charts
A,
L, T
& **U**

Although M64 is the brightest galaxy in this area, it is not the easiest to find. The dust lane that gives the illusion of a black eye is not obvious in smaller telescopes. Noted is the 5th magnitude star *35 Comae Berenices*.

For easy comparison, all photographs are shown at the same scale, measuring 2.3° x 1.3°.

M 65 ✧ Galaxy

Original Messier Description

from the 1784 edition of *Connaissance des Temps*

Observed 1780. March 1

Nebula discovered in Leo: it's very faint & contains no stars.

NGC Summary by J.L.E. Dreyer, circa 1888

NGC Number: 3623
Mag: 9.3   **Size:** 10'
NGC Description: Bright, very large, much extended approximately along position angle 165°, gradually becomes brighter in the middle to a bright nucleus.

Facts

Name: No common name
Type of Object: Spiral galaxy
Magnitude: 9.3
Distance: 29,000,000 ly
Physical Size: 84,000 ly in diameter
Arc Degree Size in Sky: 10' x 3'
Other: Shape classified as Sb, which means that it has a round nucleus with medium wound arms (see page 340).

Description of M65 using a 4-inch refractor at 48x

A pretty sight of two bright galaxies. M65 is slightly fainter and more edge on than M66. The centers of these galaxies are starlike but M66's center is more elongated. Although both galaxies are similar in length, M65 appears slightly longer.

Locating Index: Fairly easy because M65/M66 are about 2½° below the 3rd magnitude star *Chertan* in Leo.

Identifying/Observing Index: Easy because the M65/M66 galaxy pair are close together and bright, so they are easy to spot. They do fade in light-polluted skies.

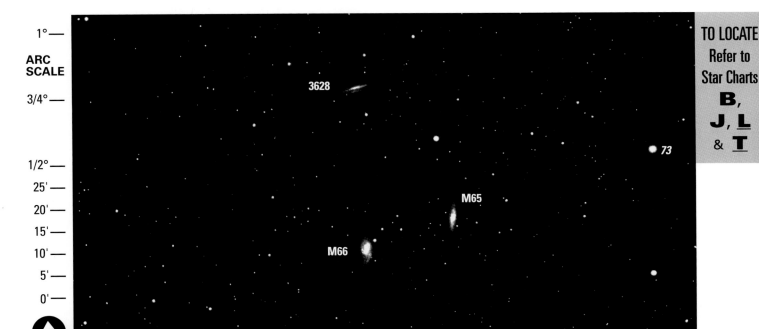

ARC SCALE

1° — 3/4° — 1/2° — 25' — 20' — 15' — 10' — 5' — 0' —

3628

M65

M66

73

TO LOCATE
Refer to
Star Charts
B,
J, L
& **T**

M65, M66 and NGC 3628 are known as the "Leo Triplet." The 5th magnitude star *73 Leonis* (Flamsteed designation) is noted.

For easy comparison, all photographs are shown at the same scale, measuring 2.3° x 1.3°.

M66 ✦ Galaxy

Original Messier Description

from the 1784 edition of *Connaissance des Temps*

Observed 1780. March 1

Nebula discovered in Leo; its light very faint, & very close to the preceding one: they both appear in the same telescopic field. The Comet observed in 1773 & 1774 had passed between these two nebulae from Nov. 1 to 2, 1773. Mr. Messier didn't see them at that time, no doubt because of the Comet's light.

NGC Summary by J.L.E. Dreyer, circa 1888

NGC Number: 3627
Mag: 8.7   **Size:** 9'
NGC Description: Bright, very large, much extended along position angle 150°, much brighter in the middle, 2 stars north preceding.

Author's Note: The word "preceding" in this NGC description refers to stars west of the galaxy and is in reference to the order in which objects move across the sky. These two stars have a Right Ascension earlier than that of the galaxy.

Location

Constellation: Leo (Leo)

Year 2000 Coordinates
RA: 11h 20.2m
Dec: +12° 59'

Observation Periods

Evenings 8 p.m. : February to July
Mornings 4 a.m. : October to March

Facts

Name: No common name
Type of Object: Spiral galaxy
Magnitude: 9.0
Distance: Estimates range from 25,000,000 ly to 39,100,000 ly
Physical Size: 65,000 ly in diameter for the closest distance estimate and 102,000 ly for the farthest
Arc Degree Size in Sky: 9' x 4'
Other: Shape classified as Sb, which means that it has a round nucleus with medium wound arms.

Description of M66 using a 4-inch refractor at 48x

You cannot describe M66 without comparing it to M65. M66 is slightly brighter and more face on than M65. The centers of these galaxies are starlike but M66's center is more elongated. Although both galaxies are similar in length, M65 appears slightly longer.

Locating Index: **Fairly easy** because M65/M66 are about 2½° below the 3rd magnitude star *Chertan* in Leo.

Identifying/Observing Index: **Easy** because the M65/M66 galaxy pair are close together and bright, so they are easy to spot. They do fade in light-polluted skies.

TO LOCATE
Refer to
Star Charts
**B,
J, L**
& **T**

M65 and M66 are favorites because they are bright and close together, so they make a spectacular sight in any eyepiece. Check out the M65 picture for additional information.

For easy comparison, all photographs are shown at the same scale, measuring 2.3° x 1.3°.

M 67 King Cobra ✧ Open Cluster

Original Messier Description

from the 1784 edition of *Connaissance des Temps*

Observed 1780. Apr. 6

Cluster of small stars with nebulosity below the southern Claw of Cancer. The position determined from the star α.

Location

Constellation: Cancer (Cnc)

<u>**Year 2000 Coordinates**</u>
RA: 8h 51.4m
Dec: +11° 49'

Observation Periods

Evenings 8 p.m. : January to June
Mornings 4 a.m. : October to February

Facts

Name: King Cobra
Type of Object: Open cluster
Magnitude: 6.9
Distance: 2,700 ly
Physical Size: Spans 24 ly
Arc Degree Size in Sky: Extends 30'
Other: Contains about 200 stars, the brightest shining at magnitude 9.7. This is an old cluster with an age estimated at 3,200,000,000 years.

NGC Summary by J.L.E. Dreyer, circa 1888

NGC Number: 2682
Mag: 6.9   **Size:** 30'
NGC Description: Remarkable! Cluster, very bright, very large, extremely rich in stars, little compressed, contains stars of magnitude 10 to 15.

Description of M67 using a 4-inch refractor at 48x

Composed of fainter stars with one brighter, that stands out. The majority of its stars fit in a semicircle with the brightest star lying outside in a little curved band "tangent" to the semicircle. Many more stars are visible with averted vision.

Locating Index: Slightly challenging because it is not near any conspicuously bright stars. However, this cluster forms the apex of a shallow isosceles triangle with the stars *Regulus* and *Procyon* placing it slightly above the halfway point between these two bright stars.

Identifying/Observing Index: Fairly easy because it is in an area where stars are sparse so it "pops" when you come across it.

M67

TO LOCATE
Refer to
Star Charts
B &
J

M67 is a beautiful cluster. The numbers 50 and 60 in this picture are the Flamsteed designations of the 5½ magnitude star *50 Cancri* and 4th magnitude star *60 Cancri*.

For easy comparison, all photographs are shown at the same scale, measuring 2.3° x 1.3°.

M68 ✧ Globular Cluster

Original Messier Description

from the 1784 edition of *Connaissance des Temps*

Observed 1780. Apr. 9

Nebula without stars below Corvus & Hydra; it's very faint, very hard to see with the telescope; there is a sixth magnitude star close to it.

Location

Constellation: Hydra (Hya)

Year 2000 Coordinates
RA: 12h 39.5m
Dec: −26° 45'

Observation Periods

Evenings 8 p.m. : April to July
Mornings 4 a.m. : December to March

Facts

Name: No common name
Type of Object: Globular cluster
Magnitude: 8.2
Distance: 33,000 ly
Physical Size: 115 ly in diameter
Arc Degree Size in Sky: 12' in diameter
Other: The stars in this globular cluster are much more loosely compacted than average.

NGC Summary by J.L.E. Dreyer, circa 1888

NGC Number: 4590
Mag: 8.2   **Size:** 12'
NGC Description: Globular cluster of stars, large, extremely rich in stars, very compressed, irregularly round, well resolved, contains stars of magnitude 12.

Description of M68 using a 4-inch refractor at 48x

Located very far south, in an area devoid of stars. It is similar to M53 that lies farther north but with only about half its overall brightness. The center part of its core is brightest, then it gradually fades outward. Would this appear brighter if it were higher in the sky?

Locating Index: **Slightly challenging.**

This cluster is almost on the line formed by the stars δ and β *Corvi* (Corvus), at about half the distance between them, south and beyond β.

Identifying/Observing Index: **Fairly easy** in darker skies, but it will prove **more challenging** in those skies with light pollution.

TO LOCATE
Refer to
Star Charts
A,
L
& N

ARC SCALE

1°
3/4°
1/2°
25'
20'
15'
10'
5'
0'

A

M68, centered, appears as a fainter version of M53 farther north. The 5th magnitude star marked with an A became bloated and elongated when a wispy cloud strand passed in front of it.

For easy comparison, all photographs are shown at the same scale, measuring 2.3° x 1.3°.

M69 ✧ Globular Cluster

Original Messier Description

from the 1784 edition of *Connaissance des Temps*

Observed 1780. August 31

Nebula without stars, in Sagittarius, below his left arm & near the bow. Close to it, there is a 9th magnitude star; its light is very faint, it can only be seen in good weather, & the slightest light used to illuminate the micrometer wires made it disappear. Its position was determined from ε of Sagittarius. This nebula has been observed by Mr. de la Caille & reported in his Catalogue. It resembles the nucleus of a little Comet.

NGC Summary by J.L.E. Dreyer, circa 1888

NGC Number: 6637
Mag: 7.7 **Size:** 7.1'
NGC Description: Globular cluster of stars, bright, large, round, well resolved, contains stars of magnitude 14 to 16.

Location

Constellation: Sagittarius (Sgr)

Year 2000 Coordinates
RA: 18h 31.4m
Dec: −32° 21'

Observation Periods

Evenings 8 p.m. : August to October
Mornings 4 a.m. : April to June

Facts

Name: No common name
Type of Object: Globular cluster
Magnitude: 7.6
Distance: 28,000 ly
Physical Size: 58 ly in diameter
Arc Degree Size in Sky: 7.1' in diameter
Other: The compactness of the stars in this globular cluster is slightly tighter than average.

Description of M69 using a 4-inch refractor at 48x

M69 is about as bright as M54 but it has more fuzziness. Its center is starlike with a surrounding glow. About the same size as M70, but brighter.

Locating Index: Fairly easy because it forms the apex of an isosceles triangle on the quarter line between the stars ζ *Sagittarii* (*Ascella*) and *Kaus Australis*.

Identifying/Observing Index: Fairly easy in dark skies. It will prove a little **challenging** in light-polluted skies because it is not very bright. It's easier to spot than M70.

TO LOCATE Refer to Star Charts **A**, **P** & **S**

M69 boasts an 8th magnitude star 5' away. This globular is also near a string of 5th and 7th magnitude stars less than a degree south of it where the much fainter globular NGC 6652 is located.

For easy comparison, all photographs are shown at the same scale, measuring 2.3° x 1.3°.

M70 ✧ Globular Cluster

Original Messier Description

from the 1784 edition of *Connaissance des Temps*

Observed 1780. August 31

Nebula without stars, near the preceding one & on the same parallel. Close to it are a ninth magnitude star & four small telescopic stars, almost along the same straight line, very close to each other & located above the nebula, as seen in a reversing telescope. The nebula determined from the same star, ε of Sagittarius.

NGC Summary by J.L.E. Dreyer, circa 1888

NGC Number: 6681
Mag: 8.1   **Size:** 7.8'
NGC Description: Globular cluster of stars, bright, pretty large, round, gradually becomes brighter in the middle, contains stars of magnitude 14 to 17.

Location

Constellation: Sagittarius (Sgr)
Year 2000 Coordinates
RA: 18h 43.2m
Dec: –32° 18'

Observation Periods

Evenings 8 p.m. : August to October
Mornings 4 a.m. : April to June

Facts

Name: No common name
Type of Object: Globular cluster
Magnitude: 8.1
Distance: 29,000 ly
Physical Size: 67 ly in diameter
Arc Degree Size in Sky: 8' in diameter
Other: The compactness of the stars in this globular cluster is slightly tighter than average and is the same as that of M69.

Description of M70 using a 4-inch refractor at 48x

A scaled down duplicate of M54 that is slightly smaller and fainter. Its somewhat starlike center is half as bright as M54's.

Locating Index: Easy because it is halfway between the bright stars ζ *Sagittarii* (*Ascella*) and *Kaus Australis* that make up Sagittarius.

Identifying/Observing Index: Fairly easy in dark skies. It will prove more **challenging** in light-polluted skies because it is not very bright. This is the faintest of the three globulars below Sagittarius.

ARC SCALE

1°—
3/4°—
1/2°—
25'—
20'—
15'—
10'—
5'—
0'—

TO LOCATE
Refer to
Star Charts
A,
P
& **S**

Because of its location, M70, centered, is easy to find. Note the string of 9th magnitude stars 10' south of the globular.

For easy comparison, all photographs are shown at the same scale, measuring 2.3° x 1.3°.

M71 ✦ Globular Cluster

Original Messier Description

from the 1784 edition of *Connaissance des Temps*

Observed 1780. Oct. 4

Nebula discovered by Mr. Méchain on June 28, 1780, between the stars γ & δ of Sagitta. The following October 4, Mr. Messier searched for it. Its light is very faint & contains no stars. The slightest light made it disappear. It's located about 4 degrees below the one that Mr. Messier discovered in Vulpecula. See *No. 27*. He reports it on the Chart of the Comet of 1779.

Author's Note: See note for M58 about "the slightest light made it disappear."

NGC Summary by J.L.E. Dreyer, circa 1888

NGC Number: 6838
Mag: 8.3 **Size:** 7.2'
NGC Description: Cluster, very large, very rich in stars, pretty much compressed, contains stars of magnitude 11 to 16.

Location

Constellation: Sagitta (Sge)
Year 2000 Coordinates
RA: 19h 53.8m
Dec: +18° 47'

Observation Periods

Evenings 8 p.m. : June to December
Mornings 4 a.m. : February to August

Facts

Name: No common name
Type of Object: Globular cluster
Magnitude: 8.2
Distance: 13,000 ly
Physical Size: 27 ly in diameter
Arc Degree Size in Sky: 7.2' in diameter
Other: The compactness of the stars in this globular cluster has, ironically, never been rated because it was originally thought to be an open cluster. Therefore, the compactness is probably extremely loose.

Description of M71 using a 4-inch refractor at 48x

A "fuzzy" globular with an evenly lighted and diffused center. Medium in size and not as bright as you might wish. I cannot see individual stars. There is a clump of stars next to this globular that almost looks like it could be an NGC cluster.

Locating Index: Fairly easy as long as you can see the two 3½ magnitude stars, γ and δ *Sagittae*, because the globular lies midway between them but slightly south of the line.

Identifying/Observing Index: Slightly challenging because it is not very bright and blends in with the Milky Way. Look for a subtle increase in brightness compared to the surrounding stars.

TO LOCATE
Refer to
Star Charts
A,
P
& **R**

ARC SCALE

1° —
3/4° —
1/2° —
25' —
20' —
15' —
10' —
5' —
0' —

H20

M71, centered, is in a thick part of the Milky Way. Just 25' away, there is a small open cluster designated Harvard 20, but there are also many other clumps of stars in the area.

For easy comparison, all photographs are shown at the same scale, measuring 2.3° x 1.3°.

M72 ✧ Globular Cluster

Original Messier Description

from the 1784 edition of *Connaissance des Temps*

Observed 1780. Oct. 4

Nebula seen by Mr. Méchain on the night of August 29 to 30, 1780, above the neck of Capricornus. Mr. Messier searched for it on the following October 4 & 5: its light faint like the preceding one. Near it there's a little telescopic star: its position was determined from the star γ of Aquarius, fifth magnitude.

Location

Constellation: Aquarius (Aqr)

Year 2000 Coordinates
RA: 20h 53.5m
Dec: −12° 32'

Observation Periods

Evenings 8 p.m. : August to November
Mornings 4 a.m. : April to July

Facts

Name: No common name
Type of Object: Globular cluster
Magnitude: 9.3
Distance: 55,000 ly
Physical Size: 96 ly in diameter
Arc Degree Size in Sky: 6' in diameter
Other: The compactness of the stars in this globular cluster is very loose compared to the average.

NGC Summary by J.L.E. Dreyer, circa 1888

NGC Number: 6981
Mag: 9.4 **Size:** 5.9'
NGC Description: Globular cluster of stars, pretty bright, pretty large, round, gradually becomes much more compressed in the middle, well resolved.

Description of M72 using a 4-inch refractor at 48x

Fairly small with light spread thinly over the area, making its overall surface brightness very faint, similar to the effect of the face-on spiral M33. Its center is somewhat starlike but faint.

Locating Index: **Challenging** because there are no conspicuously bright stars in the area. It approximately forms the right angle of a triangle with the stars *θ Capricorni* and *Dabih*.

Identifying/Observing Index: One of the more **challenging** Messier objects because it is faint. You need dark skies, otherwise it will be difficult to see.

TO LOCATE
Refer to
Star Charts
A,
P
& **R**

Of the 110 objects, M72 stands out as one of the more difficult to see. It is nowhere near as bright visually as in this picture, which makes it look as bright as the 8th magnitude star 15' above it.

For easy comparison, all photographs are shown at the same scale, measuring 2.3° x 1.3°.

M 73 ✧ Triangular "Asterism" of Four Stars

Original Messier Description

from the 1784 edition of *Connaissance des Temps*

Observed 1780. Oct. 4 & 5

Cluster of three or four small stars which looks like a nebula at first sight, contains a little nebulosity: this cluster is located on the parallel of the preceding nebula: its position was determined from the same star, γ of Aquarius.

Author's Note: M73 is Messier's only "mistake" or misidentification of a deep sky object. See my note for the NGC Summary below.

NGC Summary   by J.L.E. Dreyer, circa 1888

NGC Number: 6994
Mag: 9p   **Size:** 3'
NGC Description: Cluster, extremely sparse in stars, very little compressed, no nebulosity.

Author's Note: The "p" after the magnitude indicates the blue (color) photographic magnitude. In reading this NGC Description, one certainly gets the impression that it was in reaction to the original Messier description.

Location

Constellation: Aquarius (Aqr)
Year 2000 Coordinates
RA: 20h 58.9m
Dec: −12° 38'

Observation Periods

Evenings 8 p.m. : August to November
Mornings 4 a.m. : April to July

Facts

Name: No common name
Type of Object: Four stars of magnitude 10½ or fainter forming a small equilateral triangle
Magnitude: Brightest around 10½
Distance: Unknown for any of these stars
Physical Size: Not applicable
Arc Degree Size in Sky: Each side of the triangle measures approximately 0.7'
Other: Could it have been Messier's telescope optics that created something resembling a nebula with this triangle of stars?

Description of M73 using a 4-inch refractor at 48x

On casual inspection, it looks like three faint stars forming the shape of a triangle. However, with averted vision, I can just make out a fourth star that is close to one of the others.

Locating Index: Challenging because there are no conspicuously bright stars in the area. It is easiest to find the challenging M72 and move east a little more than a degree until you spot it.

Identifying/Observing Index: Fairly easy to spot this triangle of stars because it does stand out as an interesting little asterism or arrangement.

TO LOCATE
Refer to
Star Charts
A,
P
& **R**

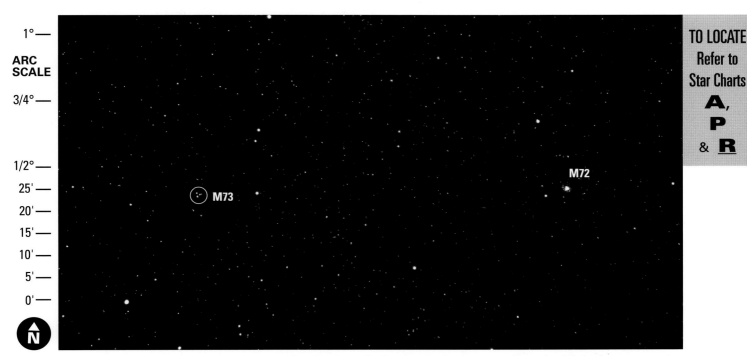

This is the only object Messier mistakenly thought had nebulosity associated with it. It may not even be an open cluster, even though it has an NGC number.

For easy comparison, all photographs are shown at the same scale, measuring 2.3° x 1.3°.

M 74 The Phantom ✧ Galaxy

Original Messier Description

from the 1784 edition of *Connaissance des Temps*

Observed 1780. Oct. 18

Nebula without stars, near the star η of the Link of Pisces, seen by Mr. Méchain at the end of September 1780, & he reports it. "This nebula doesn't contain any stars; it's fairly large, very obscure, and extremely difficult to observe; it will be possible to determine it more exactly during the good frosts." Mr. Messier searched for it & found it, as Mr. Méchain describes: it has been directly compared with the star η of Pisces.

NGC Summary by J.L.E. Dreyer, circa 1888

NGC Number: 628
Mag: 9.2   **Size:** 10.2'
NGC Description: Globular cluster of stars, faint, very large, round, very gradually then pretty suddenly becomes much brighter in the middle, partially resolved. *See Author's Note to right.*

Location

Constellation: Pisces (Psc)

Year 2000 Coordinates
RA: 1h 36.7m
Dec: +15° 47'

Observation Periods

Evenings 8 p.m. : September to March
Mornings 4 a.m. : May to November

Facts

Name: The Phantom
Type of Object: Spiral galaxy
Magnitude: 9.2
Distance: Estimates range from 26,000,000 ly to 55,400,000 ly
Physical Size: 76,000 ly in diameter for the closest distance estimate and 161,000 ly for the farthest
Arc Degree Size in Sky: 10' x 9'
Other: Shape classified as Sc, which means that it has a round nucleus with loosely wound arms.

NGC Author's Note: The NGC description incorrectly identifies this object as a globular cluster instead of a "nebula" or galaxy.

Description of M74 using a 4-inch refractor at 48x

Extremely faint object, about the size of M1. It is at my "edge" of perceptibility with the 4-inch telescope. It looks like a small circular area, emitting a very faint glow. Using averted vision helps to see it. M74 is the early evening object that sometimes "gets away" from being observed during a Messier Marathon.

Locating Index: Easy because it is about 1⅓° "northeast" of the 3½ magnitude star η Piscium. Also, the two brightest stars of Aries point to it.

Identifying/Observing Index: Challenging to **difficult** because it is very faint. It needs dark skies since it is easily washed out with any amount of light pollution.

ARC SCALE

1° —
3/4° —
1/2° —
25' —
20' —
15' —
10' —
5' —
0' —

M74

η

N

TO LOCATE
Refer to
Star Charts
B,
D
& **F**

Face-on spirals like M33 and M74 are very faint because their light is spread over the largest possible area. There are many NGC galaxies that are much brighter than M74.

For easy comparison, all photographs are shown at the same scale, measuring 2.3° x 1.3°.

M75 ✧ Globular Cluster

Original Messier Description

from the 1784 edition of *Connaissance des Temps*

Observed 1780. Oct. 18

Nebula without stars, between Sagittarius & the head of Capricornus; seen by Mr. Méchain on August 27 & 28, 1780. Mr. Messier searched for it on the following October 5, & on October 18 he compared it with the star No. 4, sixth magnitude, of Capricornus, according to Flamsteed. To Mr. Messier, it seemed to only be composed of very small stars, containing some nebulosity. Mr. Messier reported it as a nebula without stars. Mr. Messier saw it on October 5, but the moon was above the horizon, & it wasn't until the 18th of the same month that he was able to assess its appearance & determine its location.

NGC Summary by J.L.E. Dreyer, circa 1888

NGC Number: 6864
Mag: 8.6   Size: 6.0'
NGC Description: Globular cluster of stars, bright, pretty large, round, very much becomes brighter in the middle to a bright nucleus, partially resolved.

Location

Constellation: Sagittarius (Sgr)
Year 2000 Coordinates
RA: 20h 06.1m
Dec: −21° 55'

Observation Periods

Evenings 8 p.m. : August to November
Mornings 4 a.m. : April to July

Facts

Name: No common name
Type of Object: Globular cluster
Magnitude: 8½
Distance: 61,000 ly
Physical Size: 106 ly in diameter
Arc Degree Size in Sky: 6' in diameter
Other: The compactness of this globular cluster is extremely tight and is classified as a I on a I to XII scale.

Description of M75 using a 4-inch refractor at 48x

Overall, this globular is small and looks like a fuzzy star since its center is starlike. The brightness of the center matches that of an 8th magnitude star sitting north of it, just ¼ a degree away.

Locating Index: Challenging since it is not located near any conspicuously bright stars. Almost on the line formed by the stars *Nunki* of Sagittarius and *Deneb Algedi* of Capricornus. You will definitely have to move the telescope around and "bump" into it to find it.

Identifying/Observing Index: Fairly easy in dark skies but it will prove **challenging** in skies with light pollution.

TO LOCATE
Refer to
Star Charts
**A,
P**
& **R**

ARC SCALE

1° —
3/4° —
1/2° —
25' —
20' —
15' —
10' —
5' —
0' —

M75, just to the left of center, is a fairly small globular located in the middle of nowhere between the bright constellation of Sagittarius and the faint constellation of Capricornus.

For easy comparison, all photographs are shown at the same scale, measuring 2.3° x 1.3°.

M76 Little Dumbbell ✧ Planetary Nebula

Original Messier Description

from the 1784 edition of *Connaissance des Temps*

Observed 1780. Oct. 21

Nebula in the right foot of Andromeda, seen by Mr. Méchain on September 5, 1780, & he reports: "This nebula contains no stars; it's small & faint." On the following October 21, Mr. Messier searched for it with his achromatic telescope & it seemed to him that it was composed only of small stars, containing nebulosity, & that the slightest light used to illuminate the micrometer wires made them disappear: the position determined from the star φ of Andromeda, fourth magnitude.

Author's Note: See M58's note about the micrometer wires.

NGC Summary by J.L.E. Dreyer, circa 1888

NGC Numbers: 650 and 651
Mag: 12p   **Size:** 4.8'
NGC Descriptions: [650] Very bright, the preceding part of a double nebula. [651] Very bright, the following part of a double nebula.

See Author's Notes to the right.

Location

Constellation: Perseus (Per)
Year 2000 Coordinates
RA: 1h 42.4m
Dec: +51° 34'

Observation Periods

Evenings 8 p.m. : September to March
Mornings 4 a.m. : May to November

Facts

Name: Little Dumbbell Nebula
Type of Object: Planetary nebula
Magnitude: 11
Distance: 8,200 ly
Physical Size: Spans 5 ly
Arc Degree Size in Sky: 2' x 1'
Other: The central star which created this nebula has a magnitude of 15.9.

NGC Author's Notes: The NGC catalogue assigns a separate number to each half or lobe of the Little Dumbbell. "Preceding" refers to the western lobe and "following" to the eastern lobe.
  The "p" after the magnitude indicates the blue (color) photographic magnitude.
  In the NGC catalogue, the values given for the Magnitude and Size of 650 and 651 are identical.

Description of M76 using a 4-inch refractor at 48x

Small and faint — smaller than M1 or M57. Appears as a faint, slightly elongated irregular nebula that is evenly illuminated. There is a conspicuously bright reddish star about 10' away.

Locating Index: Fairly easy because it is within a degree of the 3½ magnitude star φ *Persei*, which is a little over 2° from the 3rd magnitude star *51 Andromedae*.

Identifying/Observing Index: Slightly challenging because it is small and faint; however, it can be plainly seen in dark skies. It will become obscured in light-polluted skies.

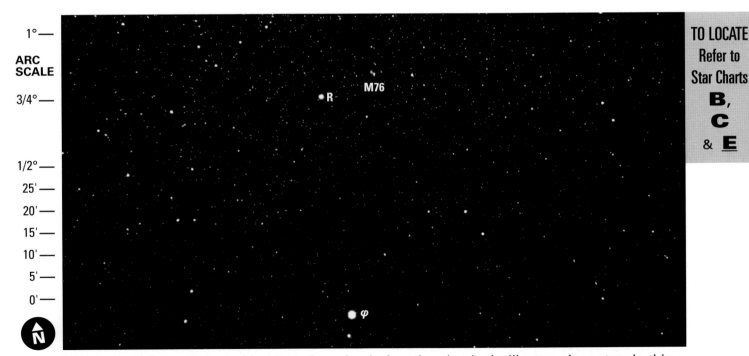

ARC SCALE

1° —
3/4° —
1/2° —
25' —
20' —
15' —
10' —
5' —
0' —

R

M76

φ

N

TO LOCATE
Refer to
Star Charts
**B,
C**
& **E**

M76 is the little two-lobed object just above its designation that looks like two close stars in this picture. Not quite a degree below is the bright 3½ magnitude star φ *Persei*. The nearby reddish star is indicated by R.

For easy comparison, all photographs are shown at the same scale, measuring 2.3° x 1.3°.

M77 ✦ Galaxy

Original Messier Description

from the 1784 edition of *Connaissance des Temps*

Observed 1780. Dec. 17

Cluster of small stars which contains some nebulosity in the Whale & on the parallel of the star δ reported to be of third magnitude, but that Mr. Messier estimates to only be of fifth. Mr. Méchain saw this cluster on October 29, 1780, in the form of a nebula.

NGC Summary   by J.L.E. Dreyer, circa 1888

NGC Number: 1068
Mag: 8.8   **Size:** 6.9'
NGC Description: Very bright, pretty large, irregularly round, suddenly becomes brighter in the middle with some stars seen near the nucleus.

Location

Constellation: Cetus (Cet)

Year 2000 Coordinates
RA: 2h 42.7m
Dec: −0° 01'

Observation Periods

Evenings 8 p.m. : October to March
Mornings 4 a.m. : June to November

Facts

Name: No common name
Type of Object: Spiral galaxy
Magnitude: 8.8
Distance: Estimates range from 52,000,000 ly to 81,500,000 ly
Physical Size: 106,000 ly in diameter for the closest distance estimate and 166,000 ly for the farthest
Arc Degree Size in Sky: 7' x 6'
Other: The shape of this galaxy is classified as Sb, meaning that it has a round nucleus with medium wound arms. It is also a Seyfert galaxy, which means that it has a small bright nucleus with the peculiar property of fluctuating in brightness.

Description of M77 using a 4-inch refractor at 48x

This spiral galaxy is similar to M74 because it is close to a bright star (δ Ceti) and difficult to see. It appears as two "stars," one shining at a faint 10th magnitude and positioned just outside the outer edge of this galaxy, while the other "fuzzy" star is its nucleus.

Locating Index: **Easy** because it is located almost 1° away from the 3½ magnitude star δ Ceti. It also forms a right triangle with δ Ceti and 84 Ceti.

Identifying/Observing Index: **Challenging.** Like most face-on spiral galaxies, it has low surface brightness, making it difficult to see. Similar to M74, so it will be difficult, if not impossible to see in light-polluted skies.

TO LOCATE
Refer to
Star Charts
B,
D
& **F**

The edge-on spiral galaxy NGC 1055 is noted at the top. M77, centered, looks more like an elliptical than a spiral galaxy and forms a nice right triangle with 3½ magnitude δ Ceti and 5th magnitude 84 Ceti. This area is riddled with galaxies. For easy comparison, all photographs are shown at the same scale, measuring 2.3° x 1.3°.

M78 ✧ Nebula

Original Messier Description

from the 1784 edition of *Connaissance des Temps*

Observed 1780. Dec. 17

Cluster of stars with a lot of nebulosity in Orion & on the parallel of the star δ in the Baldric, which was used to determine its location; the cluster follows the star on right ascension by 3^d 41' & the cluster higher than the star by 27' 7". Mr. Méchain had seen this cluster at the beginning of 1780, & reports it thus: "On the left side of Orion, at 2 to 3 minutes in diameter, you see two fairly bright nuclei, surrounded by nebulosity."

Author's Note: "Baldric" is a shoulder belt that holds or supports a sword.

NGC Summary by J.L.E. Dreyer, circa 1888

NGC Number: 2068
Mag: 8 **Size:** 8'
NGC Description: Bright, large, wisp; gradually becomes much brighter to the nucleus, 3 stars involved, mottled.

Location

Constellation: Orion (Ori)
Year 2000 Coordinates
RA: 5h 46.7m
Dec: +0° 03'

Observation Periods

Evenings 8 p.m. : December to April
Mornings 4 a.m. : August to December

Facts

Name: No common name
Type of Object: Reflection nebula
Magnitude: 8
Distance: 1,600 ly
Physical Size: Spans 4 ly
Arc Degree Size in Sky: 8' x 6'
Other: Three other small nebulae are associated with M78. The most discernible is NGC 2071 which is 15' away, appearing as a star with a little nebulosity around it.

Description of M78 using a 4-inch refractor at 48x

Looking straight on, you see nebulosity around two faint stars of magnitude 8 or 9. The nebulosity is brighter near the stars and then fades away. Its shape is conical. Not too many stars are immediately around it.

Locating Index: Fairly easy because it forms a right angle with the belt star *Alnitak* and it is as far from *Alnitak* as the width of the three belt stars.

Identifying/Observing Index: Fairly easy and plainly visible in dark skies. It will fade in light-polluted skies.

ARC SCALE
1°—
3/4°—
1/2°—
25'—
20'—
15'—
10'—
5'—
0'—

2071

TO LOCATE
Refer to
Star Charts
B,
F
& **H**

There is a 9½ magnitude star that overlaps the boundary of M78. The nebulosity around NGC 2071 can just be glimpsed. The shape of M78 reminds me of "Hubble's Variable Nebula," NGC 2261 in the constellation Monoceros.

For easy comparison, all photographs are shown at the same scale, measuring 2.3° x 1.3°.

M79 ✧ Globular Cluster

Original Messier Description

from the 1784 edition of *Connaissance des Temps*

Observed 1780. Dec. 17

Nebula without stars, situated below Lepus & on the parallel of a sixth magnitude star; seen by Mr. Méchain on October 26, 1780. Mr. Messier searched for it on the following December 17. This nebula is beautiful, the center brilliant, the nebulosity not very diffuse; its position determined from the star ε of Lepus, fourth magnitude.

Location

Constellation: Lepus (Lep)

Year 2000 Coordinates
RA: 5h 24.5m
Dec: −24° 33'

Observation Periods

Evenings 8 p.m. : January to April
Mornings 4 a.m. : September to December

Facts

Name: No common name
Type of Object: Globular cluster
Magnitude: 7.7
Distance: 42,000 ly
Physical Size: 106 ly in diameter
Arc Degree Size in Sky: 8.7' in diameter
Other: The compactness of the stars in this globular cluster is slightly tighter than average.

NGC Summary by J.L.E. Dreyer, circa 1888

NGC Number: 1904
Mag: 8.0   **Size:** 8.7'
NGC Description: Globular cluster of stars, pretty large, extremely rich in stars, extremely compressed, well resolved.

Description of M79 using a 4-inch refractor at 48x

Its center is slightly extended, looking almost starlike, then it fades beyond. A 5th magnitude double star is positioned ½° away. This globular isn't spectacular in small scopes but comes "alive" in larger ones.

Locating Index: Fairly easy because it is almost on the same line as α and β *Leporis* and about as far from β as the distance between α and β.

Identifying/Observing Index: Fairly easy in dark skies because it is fairly bright but it will prove **more challenging** in light-polluted skies.

ARC SCALE

1° —
3/4° —
1/2° —
25' —
20' —
15' —
10' —
5' —
0' —

TO LOCATE Refer to Star Charts **B, F** & **H**

M79, centered, is really the only example of a globular cluster visible during the winter.

For easy comparison, all photographs are shown at the same scale, measuring 2.3° x 1.3°.

265

M80 ✧ Globular Cluster

Original Messier Description
from the 1784 edition of *Connaissance des Temps*

Observed 1781. Jan. 4

Nebula without stars in Scorpius, between the stars g & δ; compared with g to determine its position. This nebula is round, the center is brilliant & resembles the nucleus of a little Comet, surrounded by nebulosity. Mr. Méchain saw it on January 27, 1781.

Author's Note: The star "g" is the star ρ *Scorpii*. On charts during Messier's time, Roman letters were used to designate stars after the Greek letters had been exhausted — only a few holdovers exist today.

Location
Constellation: Scorpius (Sco)
Year 2000 Coordinates
RA: 16h 17.0m
Dec: −22° 59'

Observation Periods
Evenings 8 p.m. : June to September
Mornings 4 a.m. : February to May

NGC Summary by J.L.E. Dreyer, circa 1888
NGC Number: 6093
Mag: 7.2 **Size:** 8.9'
NGC Description: Very remarkable!! Globular cluster of stars, very bright, large, very much becomes brighter in the middle (contains a variable star), well resolved, contains stars of magnitude 14.

Facts
Name: No common name
Type of Object: Globular cluster
Magnitude: 7.3
Distance: 33,000 ly
Physical Size: 86 ly in diameter
Arc Degree Size in Sky: 9' in diameter
Other: The compactness of the stars in this globular cluster is very much tighter than average.

Description of M80 using a 4-inch refractor at 48x

It has an intense center that looks starlike. When you look directly at this globular, it looks like a star and there is a star next to it with the same brightness. The fuzziness of this cluster is more apparent using averted vision. About average in size.

Locating Index: Fairly easy because it forms the apex of a shallow isosceles triangle with the stars *Antares* and δ *Scorpii*.

Identifying/Observing Index: Fairly easy because its center is intense and bright. In skies that are light-polluted, look for a slightly fuzzy star.

1° —

ARC SCALE

3/4° —

1/2° —
25' —
20' —
15' —
10' —
5' —
0' —

TO LOCATE
Refer to
Star Charts
A,
N
& **P**

M80 has a bright, starlike core matching the magnitude of an adjacent star.

For easy comparison, all photographs are shown at the same scale, measuring 2.3° x 1.3°.

M81 ✦ Galaxy

Original Messier Description

from the 1784 edition of *Connaissance des Temps*

Observed 1781. Feb. 9

Nebula near the ear of Ursa Major, on the parallel of the star d, of fourth or fifth magnitude: its position determined from that star. This nebula is slightly oval, the center clear, & you can see it very well with an ordinary three-&-a-half-foot telescope. It was discovered in Berlin by Mr. Bode on December 31, 1774, & by Mr. Méchain in August 1779.

Author's Note: Johann Elert Bode (1747–1826) was a Berlin astronomer who published a catalogue of 75 deep sky objects in 1777; however, many of these objects were copied from other catalogues or listings.

NGC Summary by J.L.E. Dreyer, circa 1888

NGC Number: 3031
Mag: 6.9   **Size:** 25.7'
NGC Description: Remarkable! Extremely bright, extremely large, extended along position angle 156°, gradually and then suddenly becomes very much brighter in the middle to a bright nucleus.

Location

Constellation: Ursa Major (UMa)
Year 2000 Coordinates
RA: 9h 55.6m
Dec: +69° 04'

Observation Periods

Evenings 8 p.m. : January to July
Mornings 4 a.m. : September to April

Facts

Name: Bode's Galaxies (M81/M82 pair)
Type of Object: Spiral galaxy
Magnitude: 6.8
Distance: 9,500,000 ly
Physical Size: 72,000 ly in diameter
Arc Degree Size in Sky: 26' x 14'
Other: The shape of this galaxy is classified as Sb, which means that it has a round nucleus with medium wound arms (see page 340).

Description of M81 using a 4-inch refractor at 48x

Both M81 and M82 appear in the same eyepiece field of view. M81 is the more circular, face-on spiral. Compared to M82, M81 is much brighter and slightly longer. It appears more diffused or nebulose and its center is starlike. A favorite.

Locating Index: Challenging because M81/M82 are in an area sparse of brighter, conspicuous stars. To locate, point the telescope at a spot on the same line as the bowl stars, *Dubhe* and *Phad* and at the same distance as these two stars are from *Dubhe* (see line on Chart I). Sweep until you come across either one.

Identifying/Observing Index: Easy because the M81/82 pair of galaxies is large and bright. They are visible even in light-polluted skies.

TO LOCATE
Refer to
Star Charts
B,
I
& **K**

ARC SCALE

1° —
3/4° —
1/2° —
25' —
20' —
15' —
10' —
5' —
0' —

N

M81 is farther south (bottom) than M82 (top). What is wonderful about these galaxies is their brightness and close proximity to one another, which allow both to be viewed in light-polluted skies and in the same eyepiece field of view. For easy comparison, all photographs are shown at the same scale, measuring 2.3° x 1.3°.

M82 Cigar Galaxy ✧ Galaxy

Original Messier Description

from the 1784 edition of *Connaissance des Temps*

Observed 1781. Feb. 9

Nebula without stars, near the preceding one; they both appear at the same time in the telescope's field, this one less visible than the preceding one; its light faint & elongated; there's a telescopic star on its end. Seen in Berlin by Mr. Bode on December 31, 1774, & by Mr. Méchain in August 1779.

Location

Constellation: Ursa Major (UMa)
Year 2000 Coordinates
RA: 9h 55.8m
Dec: +69° 41'

Observation Periods

Evenings 8 p.m. : January to July
Mornings 4 a.m. : September to April

Facts

Name: Cigar Galaxy, Bode's Galaxies (M81/M82 pair)
Type of Object: Irregular galaxy
Magnitude: 8.4
Distance: 9,500,000 ly
Physical Size: Spans 30,000 ly
Arc Degree Size in Sky: 11' x 5'
Other: The Hubble Space Telescope has taken incredibly detailed images of this galaxy which suggest that it is round and flat like a dish (similar to a spiral), but with very mixed-up insides.

Description of M82 using a 4-inch refractor at 48x

Both M81 and M82 appear in the same eyepiece field of view. M82 is the cigar-shaped galaxy. Compared to M81, M82 is much fainter and slightly shorter. It has a 10th magnitude star next to it. These two galaxies together are sometimes referred to as Bode's Nebulae. A favorite.

Locating Index: Challenging because M81/M82 are in an area sparse of brighter, conspicuous stars. To locate, see the Locating Index for M81 and the directional line drawn on Chart I.

Identifying/Observing Index: Easy because the M81/82 pair of galaxies is large and bright. They are visible even in light-polluted skies.

ARC SCALE

1° —
3/4° —
1/2° —
25' —
20' —
15' —
10' —
5' —
0' —

N

3077

TO LOCATE
Refer to
Star Charts
B,
I
& **K**

Of the M81/M82 pair, I have a difficult time remembering that M82 is the one with the cigar shape. There is a fainter galaxy, NGC 3077, that is at a similar distance from M81 as M81 is from M82.

For easy comparison, all photographs are shown at the same scale, measuring 2.3° x 1.3°.

M83 ✦ Galaxy

Original Messier Description

from the 1784 edition of *Connaissance des Temps*

Observed 1781. Feb. 17

Nebula without stars near the head of Centaurus: it appears as a faint & even glow, but is so difficult to see in the telescope that the slightest illumination of the micrometer wires makes it disappear. It will take a lot of effort for anyone to see it: it forms a triangle with two stars, estimated at sixth & seventh magnitude. Determined from the stars i, k, h in the head of Centaurus. Mr. de la Caille had already determined this nebula. See the end of this Catalogue.

NGC Summary by J.L.E. Dreyer, circa 1888

NGC Number: 5236
Mag: 7.6   **Size:** 11.2'
NGC Description: Very remarkable!! Very bright, very large, extended along position angle 55°, becomes extremely and suddenly brighter in the middle to the nucleus, 3-branch spiral.

Location

Constellation: Hydra (Hya)
Year 2000 Coordinates
RA: 13h 37.0m
Dec: −29° 52'

Observation Periods

Evenings 8 p.m. : May to July
Mornings 4 a.m. : January to March

Facts

Name: No common name
Type of Object: Spiral galaxy
Magnitude: 8
Distance: Estimates range from 8,500,000 ly to 22,500,000 ly
Physical Size: 27,000 ly in diameter for the closest distance estimate and 72,000 ly for the farthest
Arc Degree Size in Sky: 11' x 10'
Other: The shape of this galaxy is classified as Sc, which means that it has a round nucleus with loosely wound arms.

Description of M83 using a 4-inch refractor at 48x

Very "large" in comparison to any galaxy in the Virgo Galaxy Cluster. This face-on spiral has a very starlike center. Overall, it's not very bright, but very extended and just plain nice.

Locating Index: Slightly challenging because it is not near any conspicuously bright stars. It forms the right angle of a triangle with π and γ *Hydrae*.

Identifying/Observing Index: Easy in dark skies because it is big and prominent. Look for a fuzzy star in light-polluted skies. I wonder what this galaxy would look like higher up in the sky, like from the southern hemisphere?

TO LOCATE
Refer to Star Charts **A, L & N**

1° —
ARC SCALE
3/4° —
1/2° —
25' —
20' —
15' —
10' —
5' —
0' —

M83 is a real treat after looking at all those smaller and fainter galaxies of the Virgo Galaxy Cluster farther north.

For easy comparison, all photographs are shown at the same scale, measuring 2.3° x 1.3°.

M84 ✧ Virgo Cluster Galaxy

Original Messier Description

from the 1784 edition of *Connaissance des Temps*

Observed 1781. March 18

Nebula without stars, in Virgo. Its center is somewhat bright, surrounded by a light nebulosity. Its brightness & general appearance resemble those of *Nos. 59 & 60* of this Catalogue.

Location

Constellation: Virgo (Vir)

Year 2000 Coordinates
RA: 12h 25.1m
Dec: +12° 53'

Observation Periods

Evenings 8 p.m. : March to August
Mornings 4 a.m. : November to April

Facts

Name: No common name
Type of Object: Elliptical galaxy
Magnitude: 9.3
Distance: 56,000,000 ly
Physical Size: Spans at least 81,000 ly
Arc Degree Size in Sky: 5' x 4'
Other: The shape of this galaxy is classified as E1, which means that it is slightly out of round from a sphere (see page 340). This galaxy is a member of the Virgo Galaxy Cluster.

NGC Summary by J.L.E. Dreyer, circa 1888

NGC Number: 4374
Mag: 9.3   **Size:** 5.0'
NGC Description: Very bright, pretty large, round, pretty suddenly becomes brighter in the middle, not resolved.

Description of M84 using a 4-inch refractor at 48x

M84 and M86 are only 15' apart and will appear in the same eyepiece field of view. Although they are similar in appearance, looking like faint smudges, M84 is the smaller and fainter one. Look for a third galaxy, NGC 4388, that is half as faint and forms a nice triangle with the M84/M86 pair.

Locating Index: Challenging because it is not near any conspicuously bright stars. To locate this galaxy, use Chart U, starting at the star *Vindemiatrix* and then "star hop" until you find it.

Identifying/Observing Index: Fairly easy because it is next to M86, thus making the M84/M86 pair an easier target. Both are visible in slightly light-polluted skies.

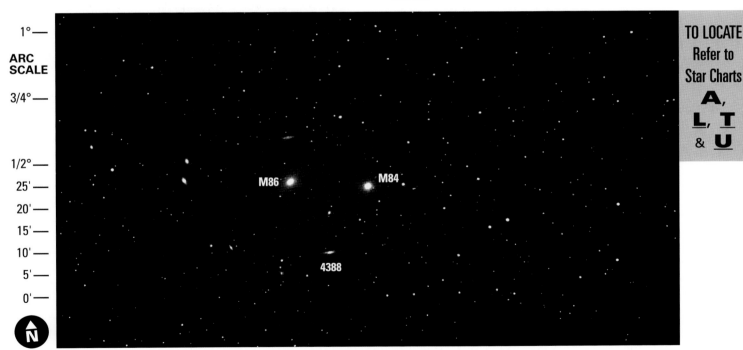

TO LOCATE
Refer to
Star Charts
A,
L, T
& **U**

Look for the fainter, edge-on spiral galaxy NGC 4388 that forms a nice equilateral triangle with M84 and M86. There are many galaxies in this area.

For easy comparison, all photographs are shown at the same scale, measuring 2.3° x 1.3°.

M85 ✦ Virgo Cluster Galaxy

Original Messier Description

from the 1784 edition of *Connaissance des Temps*

Observed 1781. March 18

Nebula without stars, above & close to the ear of Virgo, between the two stars in Coma Berenices, *Nos. 11 & 14* of Flamsteed's Catalogue: This nebula is very faint. Mr. Méchain had determined its position on March 4, 1781.

Location

Constellation: Coma Berenices (Com)

Year 2000 Coordinates
RA: 12h 25.4m
Dec: +18° 11'

Observation Periods

Evenings 8 p.m. : March to August
Mornings 4 a.m. : November to April

NGC Summary by J.L.E. Dreyer, circa 1888

NGC Number: 4382
Mag: 9.2 **Size:** 7.1'
NGC Description: Very bright, pretty large, round, brighter in the middle, star north preceding.

Author's Note: "Preceding" refers to the star positioned west of the galaxy.

Facts

Name: No common name
Type of Object: Elliptical galaxy
Magnitude: 9.2
Distance: 56,000,000 ly
Physical Size: Spans at least 114,000 ly
Arc Degree Size in Sky: 7' x 5'
Other: The shape of this elliptical galaxy is classified as Ep, where the "p" indicates that it is peculiar, thus having some oddities in its visible structure. This galaxy is a member of the Virgo Galaxy Cluster.

Description of M85 using a 4-inch refractor at 48x

This galaxy is on the fainter side, about ⅓ the brightness of M64, but not the faintest in this group of galaxies. Its center is somewhat starlike. Shows elongation. The galaxy NGC 4394, only 7' away, is as "bright" as the fainter Messier galaxies in this area.

Locating Index: Challenging because it is not near any conspicuously bright stars. To locate this galaxy, use Chart U, starting at the star *Vindemiatrix* and then "star hop" until you find it. This galaxy is about 1° from the 4½ magnitude star *11 Comae Berenices*.

Identifying/Observing Index: Slightly challenging because it is on the fainter side; however, it is visible in slightly light-polluted skies.

TO LOCATE
Refer to
Star Charts
A,
L, T
& U

ARC SCALE
1° —
3/4° —
1/2° —
25' —
20' —
15' —
10' —
5' —
0' —

4394

11

N

M85, centered, is the most northern of the Messier galaxies that make up the Virgo pack. It is also farther west but on the same Declination as the globular cluster M53. The faint galaxy NGC 4394 is noted.

For easy comparison, all photographs are shown at the same scale, measuring 2.3° x 1.3°.

M86 ✧ Virgo Cluster Galaxy

Original Messier Description

from the 1784 edition of *Connaissance des Temps*

Observed 1781. March 18

Nebula without stars in Virgo, on the same parallel as & very close to the nebula *No. 84* above: they look alike, & they both appear in the same telescopic field.

Location

Constellation: Virgo (Vir)

Year 2000 Coordinates
RA: 12h 26.2m
Dec: +12° 57'

Observation Periods

Evenings 8 p.m. : March to August
Mornings 4 a.m. : November to April

Facts

Name: No common name
Type of Object: Elliptical galaxy
Magnitude: 9.2
Distance: 56,000,000 ly
Physical Size: Spans at least 114,000 ly
Arc Degree Size in Sky: 7' x 5'
Other: The shape of this galaxy is classified as E3, which means that it is a little less elongated than "average." This galaxy is a member of the Virgo Galaxy Cluster.

NGC Summary   by J.L.E. Dreyer, circa 1888

NGC Number: 4406
Mag: 9.2   **Size:** 7.4'
NGC Description: Very bright, large, round, gradually becomes brighter in the middle to the nucleus, not resolved.

Description of M86 using a 4-inch refractor at 48x

M86 and M84 are only 15' apart, thus will appear in the same eyepiece field of view. They are similar in appearance, looking like faint smudges; however, M86 is somewhat larger and brighter. Look for a third galaxy, NGC 4388, that is half as faint and forms a nice triangle with the pair.

Locating Index: **Challenging** because it is not near any conspicuously bright stars. To locate this galaxy, use Chart U, starting at the star *Vindemiatrix* and then "star hop" until you find it.

Identifying/Observing Index: **Fairly easy** because it is next to M84 thus making the M86/M84 pair easier to identify. Both are visible in slightly light-polluted skies.

TO LOCATE
Refer to
Star Charts
A,
L, T
& **U**

ARC SCALE

1° —
3/4° —
1/2° —
25' —
20' —
15' —
10' —
5' —
0' —

M86 M84 4388

N

As you find the Messier galaxies in this area of the sky, take some time to wander and look for other galaxies. Many are visible with smaller telescopes, appearing as small, faint smudges.

For easy comparison, all photographs are shown at the same scale, measuring 2.3° x 1.3°.

M87 Virgo A ✦ Virgo Cluster Galaxy

Original Messier Description

from the 1784 edition of *Connaissance des Temps*

Observed 1781. March 18

Nebula without stars in Virgo, below & pretty close to an eighth magnitude star, the star having the same right ascension as the nebula, & its declination was 13ᵈ 42' 21" north. This nebula appears to have the same brightness as the two nebulae, *Nos. 84 & 86*.

Location

Constellation: Virgo (Vir)

Year 2000 Coordinates
RA: 12h 30.8m
Dec: +12° 24'

Observation Periods

Evenings 8 p.m. : March to August
Mornings 4 a.m. : November to April

NGC Summary by J.L.E. Dreyer, circa 1888

NGC Number: 4486
Mag: 8.6   **Size:** 7.2'
NGC Description: Very bright, very large, round, much brighter in the middle, 3rd of 3.

Author's Note: "3rd of 3" refers to the last of three galaxies and is a "note" referencing a group of three close galaxies. The other two galaxies are much fainter and smaller than M87. They are NGC 4476 (1st of 3) and NGC 4478 (2nd of 3), both next to one another and about 10 arc minutes away from the center of M87.

Facts

Name: Virgo A
Type of Object: Elliptical galaxy
Magnitude: 8.6
Distance: 56,000,000 ly
Physical Size: Spans at least 114,000 ly
Arc Degree Size in Sky: 7' in diameter
Other: This elliptical galaxy is classified as an E1, which means that its shape is slightly out of round from a sphere. M87 is also known as "Virgo A," because it is the "brightest" radio source within the constellation. At its nucleus is a supermassive black hole responsible for creating the radio noise. This galaxy is the largest galaxy in the Virgo Galaxy Cluster.

Description of M87 using a 4-inch refractor at 48x

Brighter than most of the other galaxies in this group. Core appears somewhat starlike, with nebulosity surrounding it. Several of the fainter and much smaller galaxies immediately next to M87 can be glimpsed.

Locating Index: Challenging because it is not near any conspicuously bright stars. To locate this galaxy, use Chart U, starting at the star *Vindemiatrix* and then "star hop" until you find it. About a degree away from the M84/M86 pair.

Identifying/Observing Index: Slightly challenging because it is only of average brightness for this group. Can be seen in slightly light-polluted skies.

TO LOCATE
Refer to
Star Charts
**A,
L, T
& U**

ARC SCALE
1° —
3/4° —
1/2° —
25' —
20' —
15' —
10' —
5' —
0' —

N

M87, centered, is a good example of an almost spherical elliptical galaxy. Higher-resolution pictures of this galaxy can often be found in astronomy books. This galaxy has a 4,000 ly long jet, created by a central black hole with matter spiraling inward. For easy comparison, all photographs are shown at the same scale, measuring 2.3° x 1.3°.

M88 ✧ Virgo Cluster Galaxy

Original Messier Description

from the 1784 edition of *Connaissance des Temps*

Observed 1781. March 18

Nebula without stars in Virgo, between two small stars & one sixth magnitude star which appear at the same time as the nebula in the telescope's field. It's one of the dimmest ones & resembles the one reported in Virgo, *No. 58.*

Location

Constellation: Coma Berenices (Com)

Year 2000 Coordinates
RA: 12h 32.0m
Dec: +14° 25'

Observation Periods

Evenings 8 p.m. : March to August
Mornings 4 a.m. : November to April

Facts

Name: No common name
Type of Object: Spiral galaxy
Magnitude: 9½
Distance: 56,000,000 ly
Physical Size: 114,000 ly in diameter
Arc Degree Size in Sky: 7' x 4'
Other: The shape of this galaxy is classified as Sb, which means that it has a round nucleus with medium wound arms. This galaxy is a member of the Virgo Galaxy Cluster.

NGC Summary by J.L.E. Dreyer, circa 1888

NGC Number: 4501
Mag: 9.5 **Size:** 6.9'
NGC Description: Bright, very large, very much extended.

Description of M88 using a 4-inch refractor at 48x

M88 is about average in brightness for this group of galaxies but bigger than most of them. It is also brighter and bigger than M91 a degree away. Center appears elongated, which means that this spiral galaxy is tilted to our line of sight.

Locating Index: Challenging because it is not near any conspicuously bright stars. To locate this galaxy, use Chart U, starting at the star *Vindemiatrix* and then "star hop" until you find it.

Identifying/Observing Index: Fairly easy because it is within 1° of M91 making it easier to identify. M88's brightness is average for this group of galaxies, but its size is larger.

M88 is centered in the picture, with M91 less than 1° to the east.

TO LOCATE
Refer to
Star Charts
A,
L, T
& **U**

For easy comparison, all photographs are shown at the same scale, measuring 2.3° x 1.3°.

M89 ✦ Virgo Cluster Galaxy

Original Messier Description

from the 1784 edition of *Connaissance des Temps*

Observed 1781. March 18

Nebula without stars in Virgo, not far from & on the parallel of the nebula reported above, *No. 87*. Its light was extremely faint & sparse, & it can't be seen without some effort.

NGC Summary by J.L.E. Dreyer, circa 1888

NGC Number: 4552
Mag: 9.8   **Size:** 4.2'
NGC Description: Pretty bright, pretty small in angular size, round, gradually becomes much brighter in the middle.

Location

Constellation: Virgo (Vir)

Year 2000 Coordinates
RA: 12h 35.7m
Dec: +12° 33'

Observation Periods

Evenings 8 p.m. : March to August
Mornings 4 a.m. : November to April

Facts

Name: No common name
Type of Object: Elliptical galaxy
Magnitude: 9.8
Distance: 56,000,000 ly
Physical Size: Spans at least 65,000 ly
Arc Degree Size in Sky: 4' in diameter
Other: This elliptical galaxy is classified as E0, which means that it may be shaped like a perfect sphere, or that its orientation in space to our line of sight presents a perfect sphere. It is a member of the Virgo Galaxy Cluster.

Description of M89 using a 4-inch refractor at 48x

Very starlike, which is what you would expect from an "E0" elliptical galaxy. Brighter than M90 about ½° away, but much smaller; in fact, visually it appears much smaller than the Ring Nebula, M57.

Locating Index: Challenging because it is not near any conspicuously bright stars. To locate this galaxy, use Chart U, starting at the star *Vindemiatrix* and then "star hop" until you find it.

Identifying/Observing Index: Challenging because it looks more like a star than a galaxy. Can be seen in slightly light-polluted skies.

TO LOCATE
Refer to
Star Charts
A,
L, T
& **U**

M89 looks starlike because it is small and spherical in shape. Even on the original negative, this galaxy blends in with the stars around it. For easy comparison, all photographs are shown at the same scale, measuring 2.3° x 1.3°.

M90 ✧ Virgo Cluster Galaxy

Original Messier Description

from the 1784 edition of *Connaissance des Temps*

Observed 1781. March 18

Nebula without stars in Virgo. Its light is as faint as the preceding one's, *No. 89*.

Location

Constellation: Virgo (Vir)

Year 2000 Coordinates
RA: 12h 36.8m
Dec: +13° 10'

Observation Periods

Evenings 8 p.m. : March to August
Mornings 4 a.m. : November to April

Facts

Name: No common name
Type of Object: Spiral galaxy
Magnitude: 9½
Distance: 56,000,000 ly
Physical Size: 163,000 ly in diameter
Arc Degree Size in Sky: 10' x 5'
Other: The shape of this galaxy is classified as Sb, which means that it has a round nucleus with medium wound arms. It is a member of the Virgo Galaxy Cluster.

NGC Summary by J.L.E. Dreyer, circa 1888

NGC Number: 4569
Mag: 9.5   **Size:** 9.5'
NGC Description: Pretty large, brighter in the middle to the nucleus.

Description of M90 using a 4-inch refractor at 48x

One of the fainter galaxies in this group, but not the smallest. It has a somewhat starlike center, with faint nebulosity extending on opposite sides.

Locating Index: **Challenging** because
it is not near any conspicuously bright stars. To locate this galaxy, use Chart U, starting at the star *Vindemiatrix* and then "star hop" until you find it.

Identifying/Observing Index: **Challenging**
in comparison to the other galaxies in this group because it is one of the fainter ones; however, it is possible to see this galaxy in slightly light-polluted skies.

TO LOCATE
Refer to
Star Charts
**A,
L, T
& U**

Although it may not appear faint in this picture, visually M90 is one of the fainter galaxies of the Virgo group catalogued by Messier. Note the interesting configuration of the 7th and three 9th magnitude stars near the top of the picture. For easy comparison, all photographs are shown at the same scale, measuring 2.3° x 1.3°.

M91 ✦ Virgo Cluster Galaxy

Original Messier Description

from the 1784 edition of *Connaissance des Temps*

Observed 1781. March 18

Nebula without stars in Virgo, above the preceding one, *No. 90*: its light even fainter than that of the one above.

Note: The constellation Virgo & especially the northern wing, is one of the constellations that contains the most nebulae. This catalogue contains thirteen of these which have been determined, namely Nos. 49, 58, 59, 60, 61, 84, 85, 86, 87, 88, 89, 90, & 91. All these nebulae appear to have no stars; you can only see them in a very good sky & near their meridian passage. Most of these nebulae were pointed out to me by Mr. Méchain.

Author's Notes: The above note was made by Messier.
Messier's original position for M91 was incorrect because no object matched up with his coordinates. The true identity of this object has been most perplexing because this area is rich with galaxies. Which one was Messier referring to? In 1969, W. C. Williams identified M91 as the galaxy NGC 4548, believing that Messier accidentally chose M58 instead of M89 as a reference point for determining its coordinates.

NGC Summary by J.L.E. Dreyer, circa 1888

NGC Number: 4548
Mag: 10.2   **Size:** 5.4'
NGC Description: Bright, large, little extended, little brighter in the middle.

Location

Constellation: Coma Berenices (Com)
Year 2000 Coordinates
RA: 12h 35.4m
Dec: +14° 30'

Observation Periods

Evenings 8 p.m. : March to August
Mornings 4 a.m. : November to April

Facts

Name: No common name
Type of Object: Spiral galaxy
Magnitude: 10.2
Distance: 56,000,000 ly
Physical Size: 81,000 ly in diameter
Arc Degree Size in Sky: 5' x 4'
Other: The shape of this galaxy is classified as SBb, which means that it has an extended nucleus with a "bar" or "straight arm" that goes through it, and medium wound arms. This galaxy is a member of the Virgo Galaxy Cluster.

Description of M91 using a 4-inch refractor at 48x

Visually, this appears to be the faintest galaxy in this Virgo Cluster group but it is not the smallest. Its faint nucleus is somewhat starlike and has a surrounding thin veil of nebulosity. Using averted vision makes this galaxy look more pronounced.

Locating Index: Challenging because it is not near any conspicuously bright stars. To locate this galaxy, use Chart U, starting at the star *Vindemiatrix* and then "star hop" until you find it.

Identifying/Observing Index: Challenging because it is faint and relatively small. It is, however, just visible in skies with some light pollution.

TO LOCATE
Refer to
Star Charts
A,
L, **T**
& **U**

1° —
ARC SCALE
3/4° —
1/2° —
25' —
20' —
15' —
10' —
5' —
0' —

N

M91

M88

4571

M91 is fainter and smaller than nearby M88. The even fainter galaxy NGC 4571 is indicated.

For easy comparison, all photographs are shown at the same scale, measuring 2.3° x 1.3°.

M92 ✧ Globular Cluster

Original Messier Description

from the 1784 edition of *Connaissance des Temps*

Observed 1781. March 18

A fine, visible & very bright nebula, between the knee & the left leg of Hercules; can be seen very well with a one-foot telescope. It contains no stars, its center is clear & bright, is surrounded by nebulosity, & resembles the nucleus of a large Comet. Its brightness and size make it similar to the nebula which is in the belt of Hercules. See *No. 13* of this catalogue. Its position was determined by comparing it directly with the star σ of Hercules, fourth magnitude: the nebula & the star on the same parallel.

NGC Summary by J.L.E. Dreyer, circa 1888

NGC Number: 6341
Mag: 6.5   **Size:** 11.2'
NGC Description: Globular cluster of stars, very bright, very large, extremely compressed in the middle, well resolved, contains faint stars.

Location

Constellation: Hercules (Her)

Year 2000 Coordinates
RA: 17h 17.1m
Dec: +43° 08'

Observation Periods

Evenings 8 p.m. : April to November
Mornings 4 a.m. : December to July

Facts

Name: No common name
Type of Object: Globular cluster
Magnitude: 6.4
Distance: 26,000 ly
Physical Size: 85 ly in diameter
Arc Degree Size in Sky: 11.2' in diameter
Other: The compactness of the stars in this globular cluster is tighter than average.

Description of M92 using a 4-inch refractor at 48x

This is a great globular in its own right but it is smaller and not as bright as nearby M13. However, its core is brighter and more starlike than M13. Individual stars are visible with averted vision.

Locating Index: Slightly challenging.
I have always had a problem locating this one quickly. It roughly forms the apex of a shallow isosceles triangle with the stars *ι* and *π Herculis*.

Identifying/Observing Index: Easy because it is big and bright and "pops" when you come across it.

TO LOCATE
Refer to
Star Charts
**A,
M
& O**

M92 often takes a backseat to the nearby fabulous M13, but it is a spectacular globular cluster.

For easy comparison, all photographs are shown at the same scale, measuring 2.3° x 1.3°.

M93 ✦ Open Cluster

Original Messier Description

from the 1784 edition of *Connaissance des Temps*

Observed 1781. March 20

Cluster of small stars with no nebulosity between Canis Major & the prow of Navis.

Author's Note: Navis refers to the constellation Argo Navis, The Ship of the Argonauts, which has since been divided into the three constellations Vela, Puppis and Carina. See more about this in the note for M46.

Location

Constellation: Puppis (Pup)

Year 2000 Coordinates
RA: 7h 44.6m
Dec: –23° 52'

Observation Periods

Evenings 8 p.m. : February to May
Mornings 4 a.m. : October to January

Facts

Name: No common name
Type of Object: Open cluster
Magnitude: 6
Distance: 3,600 ly
Physical Size: Spans 23 ly
Arc Degree Size in Sky: Extends for 22'
Other: Composed of about 80 stars with the brightest star shining at magnitude 8.2. Contains several red colored stars. Its age is estimated at 98,000,000 years.

NGC Summary by J.L.E. Dreyer, circa 1888

NGC Number: 2447
Mag: 6.2   **Size:** 22'
NGC Description: Cluster, large, pretty rich in stars, little compressed, contains stars of magnitude 8 to 13.

Description of M93 using a 4-inch refractor at 48x

A "side" of this cluster forms an arc, with the stars packed into the concave side. Averted vision brings out many more stars than the 30 that I can count.

Locating Index: Slightly challenging, however, if you can see the 3rd magnitude "double" star ξ *Puppis*, you can find it easily. Use the brighter 2nd magnitude star ρ *Puppis* as a visual anchor.

Identifying/Observing Index: Fairly easy to spot because it "pops" when you come across it even though it is nested in an area thick with stars.

TO LOCATE Refer to Star Charts **B**, **H** & **J**

ARC SCALE
1° —
3/4° —
1/2° —
25' —
20' —
15' —
10' —
5' —
0' —

ξ

M93

M93 is in an area rich with stars and is about 1½° northwest of the 3rd magnitude "double" star ξ *Puppis*.

For easy comparison, all photographs are shown at the same scale, measuring 2.3° x 1.3°.

M94 Croc's Eye ✧ Galaxy

Original Messier Description
from the 1784 edition of *Connaissance des Temps*

Observed 1781. March 24

Nebula without stars above Charles' heart, on the parallel of the star *No. 8*, sixth magnitude, of Canes Venatici, according to Flamsteed. Its center is bright & the nebulosity is not very diffuse. It resembles the nebula which is below Lepus, *No. 79*, but this one is more beautiful & brighter. Mr. Méchain discovered it on March 22, 1781.

Author's Note: "Charles" refers to King Charles II of England whose heart was drawn over the stars of Canes Venatici by Edmond Halley in 1725 when he became Astronomer Royal. The heart was placed at the star α *Canum Venaticorum*.

NGC Summary by J.L.E. Dreyer, circa 1888
NGC Number: 4736
Mag: 8.2 **Size:** 11'
NGC Description: Very bright, large, irregularly round, very suddenly becomes very much brighter in the middle to a bright nucleus, not resolved.

Location
Constellation: Canes Venatici (CVn)
Year 2000 Coordinates
RA: 12h 50.9m
Dec: +41° 07'

Observation Periods
Evenings 8 p.m. : March to September
Mornings 4 a.m. : November to May

Facts
Name: Croc's Eye
Type of Object: Spiral galaxy
Magnitude: 8.1
Distance: 28,000,000 ly
Physical Size: 90,000 ly in diameter
Arc Degree Size in Sky: 11' x 9'
Other: The shape of this galaxy is classified as Sbp, which means that, overall, it has a round nucleus with medium wound arms but shows some peculiarities in its structure.

Description of M94 using a 4-inch refractor at 48x

Roundish in shape, so it is probably close to being face-on. Its center appears somewhat starlike and surrounded by a glow. Slightly fainter, but much smaller than nearby M51.

Locating Index: **Slightly challenging**

because it is not near any conspicuously bright stars. It does however form the apex of a shallow isosceles triangle with the stars α and β *Canum Venaticorum*. However, these two stars are not the easiest to see.

Identifying/Observing Index: **Slightly challenging** because it is not very bright.

It's easier in dark skies.

M94

TO LOCATE
Refer to
Star Charts
A/B,
K
& **M**

About a degree north of M94, there is an interesting grouping of 8th magnitude stars that may help to identify this galaxy.

For easy comparison, all photographs are shown at the same scale, measuring 2.3° x 1.3°.

M95 ✦ Galaxy

Original Messier Description

from the 1784 edition of *Connaissance des Temps*

Observed 1781. March 24

Nebula without stars in Leo above the star l: its light is very faint.

Location

Constellation: Leo (Leo)

Year 2000 Coordinates
RA: 10h 44.0m
Dec: +11° 42'

Observation Periods

Evenings 8 p.m. : February to July
Mornings 4 a.m. : October to March

Facts

Name: No common name
Type of Object: Spiral galaxy
Magnitude: 9.7
Distance: 29,000,000 ly
Physical Size: 59,000 ly in diameter
Arc Degree Size in Sky: 7' x 5'
Other: The shape of this galaxy is classified as SBb, which means that it has an extended nucleus with a "bar" or "straight arm" that goes through it, and medium wound arms.

NGC Summary by J.L.E. Dreyer, circa 1888

NGC Number: 3351
Mag: 9.7   **Size:** 7.4'
NGC Description: Bright, large, round, pretty gradually becomes much brighter in the middle to the nucleus.

Description of M95 using a 4-inch refractor at 48x

M95 is the second brightest of the M95/M96/M105 trio. Although M95 is fainter than M96, with averted vision, it appears larger or more extended because it may be more face-on. Its center is starlike.

Locating Index: Fairly easy because it forms the apex of a shallow isosceles triangle with the bright stars *Regulus* and *Chertan*.

Identifying/Observing Index: Fairly easy because the three galaxies, M95, M96 and M105 are near one another and are relatively bright. M96 is the brightest, followed by M95.

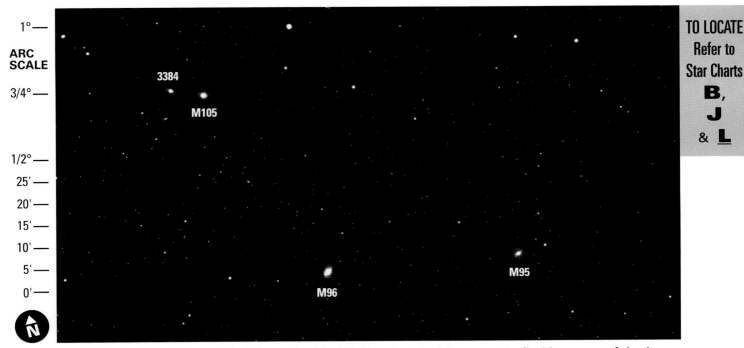

TO LOCATE
Refer to
Star Charts
B,
J
& **L**

Although this trio is not as bright as the M65/M66 pair, it is fairly easy to find because of the large area it spans in the sky — providing a greater chance of bumping into one of its galaxies. The elliptical galaxy NGC 3384 is noted.

For easy comparison, all photographs are shown at the same scale, measuring 2.3° x 1.3°.

M96 ✧ Galaxy

Original Messier Description

from the 1784 edition of *Connaissance des Temps*

Observed 1781. March 24

Nebula without stars, in Leo, near the preceding one; this one less visible, both on the parallel of *Regulus*. They resemble the two Nebulae of Virgo, *Nos. 84 & 86.* Mr. Méchain saw both of them on March 20, 1781.

NGC Summary by J.L.E. Dreyer, circa 1888

NGC Number: 3368
Mag: 9.2 **Size:** 7.1'
NGC Description: Very bright, very large, little elongated, very suddenly becomes very much brighter in the middle, not resolved.

Location

Constellation: Leo (Leo)
Year 2000 Coordinates
RA: 10h 46.8m
Dec: +11° 49'

Observation Periods

Evenings 8 p.m. : February to July
Mornings 4 a.m. : October to March

Facts

Name: No common name
Type of Object: Spiral galaxy
Magnitude: 9.2
Distance: 29,000,000 ly
Physical Size: 59,000 ly in diameter
Arc Degree Size in Sky: 7' x 5'
Other: The shape of this galaxy is classified as Sbp which means that, overall, it has a round nucleus with medium wound arms but shows some peculiarities in its structure.

Description of M96 using a 4-inch refractor at 48x

M96 is the brightest of the M95/M96/M105 trio; however, it appears slightly smaller than M95. The nucleus of M96 is starlike and similar in brightness to a star about 10' away. Appears oval in shape, so it is probably tilted a little to our line of sight.

Locating Index: Fairly easy because it forms the apex of a shallow isosceles triangle with the bright stars *Regulus* and *Chertan*.

Identifying/Observing Index: Fairly easy because the three galaxies, M95, M96 and M105 are relatively bright and near one another. M96 is the brightest, followed by M95.

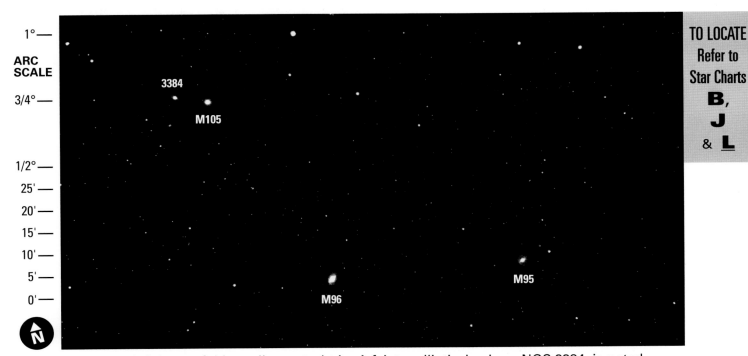

ARC SCALE

1° —
3/4° —
1/2° —
25' —
20' —
15' —
10' —
5' —
0' —

3384

M105

M96

M95

N

TO LOCATE
Refer to
Star Charts
B,
J
& **L**

M96 is the brightest of this easily spotted trio. A fainter elliptical galaxy, NGC 3384, is noted.

For easy comparison, all photographs are shown at the same scale, measuring 2.3° x 1.3°.

M97 Owl Nebula ✧ Planetary Nebula

Original Messier Description

from the 1784 edition of *Connaissance des Temps*

Observed 1781. March 24

Nebula in Ursa Major, near β. It's hard to see, as Mr. Méchain reports, especially when you illuminate the micrometer wires: its light is faint, without stars. Mr. Méchain saw it for the first time on February 16, 1781, & the position is reported according to him. Near this nebula, he saw another one that hasn't been determined yet, as well as a third one that's near γ of Ursa Major.

NGC Summary by J.L.E. Dreyer, circa 1888

NGC Number: 3587
Mag: 11.2   **Size:** 3.2'
NGC Description: Very remarkable!! Planetary nebula, very bright, very large, round, *very* gradually then very suddenly becomes brighter in the middle, 150" diameter.

Location

Constellation: Ursa Major (UMa)
Year 2000 Coordinates
RA: 11h 14.8m
Dec: +55° 01'

Observation Periods

Evenings 8 p.m. : February to August
Mornings 4 a.m. : October to April

Facts

Name: Owl Nebula
Type of Object: Planetary nebula
Magnitude: 11
Distance: Estimates range from 1,300 ly to 12,000 ly
Physical Size: 1.1 ly in diameter for the closest distance estimate and 10 ly for the farthest
Arc Degree Size in Sky: 3' in diameter
Other: This planetary, a spherical shell, is expanding at the rate of 25 miles/second or 90,000 miles/hour.

Description of M97 using a 4-inch refractor at 48x

Faint, very round and two to three times bigger than the Ring Nebula, M57. Very diffused but evenly illuminated, reminding me of a faint dime floating in the air.

Locating Index: **Easy** because it is close to the bright star *Merak* (about 2½° from it) that makes up the bottom of the bowl in the Big Dipper with *Phad*.

Identifying/Observing Index: On average, **challenging** because it is very faint. In dark skies, this planetary is fairly easy to spot, but will quickly disappear in skies with any amount of light pollution.

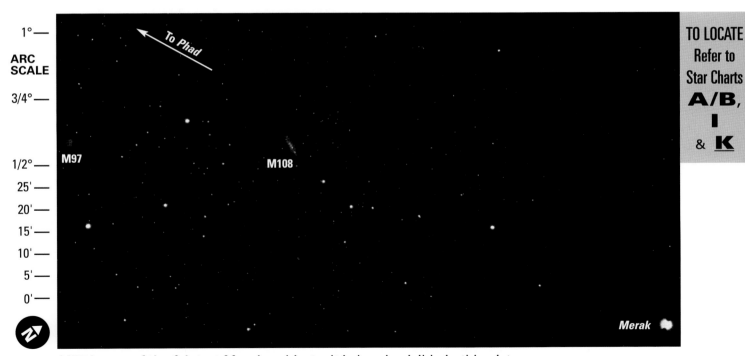

TO LOCATE
Refer to
Star Charts
A/B,
I
& **K**

M97 is one of the faintest Messier objects. It is barely visible in this picture.

For easy comparison, all photographs are shown at the same scale, measuring 2.3° x 1.3°.

M98 ✧ Virgo Cluster Galaxy

Original Messier Description

from the 1784 edition of *Connaissance des Temps*

Observed 1781. April 13

Nebula without stars, with extremely faint light, above the northern wing of Virgo; on the parallel of & close to star *No. 6*, fifth magnitude, of Coma Berenices, according to Flamsteed. Mr. Méchain saw it on March 15, 1781.

Location

Constellation: Coma Berenices (Com)

Year 2000 Coordinates
RA: 12h 13.8m
Dec: +14° 54'

Observation Periods

Evenings 8 p.m. : March to August
Mornings 4 a.m. : November to April

NGC Summary
by J.L.E. Dreyer, circa 1888
NGC Number: 4192
Mag: 10.1   **Size:** 9.5'
NGC Description: Bright, very large, very much elongated along position angle 152°, very suddenly becomes very much brighter in the middle.

Facts

Name: No common name
Type of Object: Spiral galaxy
Magnitude: 10.1
Distance: 56,000,000 ly
Physical Size: 163,000 ly in diameter
Arc Degree Size in Sky: 10' x 3'
Other: The shape of this galaxy is classified as Sb, which means that it has a round nucleus with medium wound arms. This galaxy is a member of the Virgo Galaxy Cluster.

Description of M98 using a 4-inch refractor at 48x

The only galaxy in the Virgo Galaxy Cluster group that is very edge-on and thus interesting. It's one of the fainter ones but easier to find because it is near a 5th magnitude star. It's brighter in the center and fades off to the sides. Similar to M82 but fainter.

Locating Index: Challenging because it is not near any conspicuously bright stars. To locate this galaxy, use Chart U, starting at the star *Vindemiatrix* and then "star hop" until you find it.

Identifying/Observing Index: Slightly challenging because it is not very bright. Its location near *6 Comae Berenices* helps to identify it. Can be seen in slightly light-polluted skies.

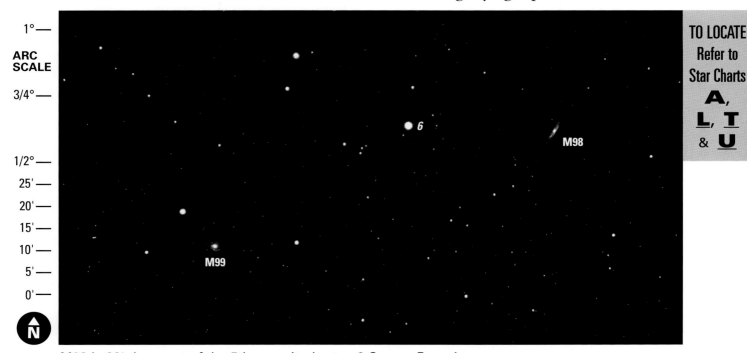

TO LOCATE
Refer to
Star Charts
**A,
L, T
& U**

M98 is 30' due west of the 5th magnitude star *6 Comae Berenices*.

For easy comparison, all photographs are shown at the same scale, measuring 2.3° x 1.3°.

M99 ✦ Virgo Cluster Galaxy

Original Messier Description

from the 1784 edition of *Connaissance des Temps*

Observed 1781. April 13

Nebula without stars, with very scarce light, but a little clearer than the preceding one, located on the northern wing of Virgo & near the same star, *No. 6*, fifth magnitude, of Coma Berenices. The nebula is between two stars, of seventh & eighth magnitude. Mr. Méchain saw it on March 15, 1781.

Location

Constellation: Coma Berenices (Com)

Year 2000 Coordinates
RA: 12h 18.8m
Dec: +14° 25'

Observation Periods

Evenings 8 p.m. : March to August
Mornings 4 a.m. : November to April

Facts

Name: Like M33 and M101, M99 is also referred to as a "Pinwheel Galaxy"
Type of Object: Spiral galaxy
Magnitude: 9.8
Distance: 56,000,000 ly
Physical Size: 81,000 ly in diameter
Arc Degree Size in Sky: 5' in diameter
Other: The shape of this galaxy is classified as Sc, which means that it has a round nucleus with loosely wound arms. This galaxy is a member of the Virgo Galaxy Cluster.

NGC Summary by J.L.E. Dreyer, circa 1888

NGC Number: 4254
Mag: 9.8   **Size:** 5.4'
NGC Description: Very remarkable!! Bright, large, round, gradually becomes brighter in the middle, not resolved, 3-branched spiral.

Description of M99 using a 4-inch refractor at 48x

Very much face on — reminding me of the Owl Nebula, M97. Its brighter center appears starlike with averted vision.

Locating Index: Challenging because it is not near any conspicuously bright stars. To locate this galaxy, use Chart U, starting at the star *Vindemiatrix* and then "star hop" until you find it.

Identifying/Observing Index: Slightly challenging because it is not very bright. Its location near *6 Comae Berenices* helps to identify it. Can be seen in slightly light-polluted skies.

TO LOCATE
Refer to
Star Charts
A,
L, T
& **U**

1°—
ARC
SCALE
3/4°—
6
M98
1/2°—
25'—
20'—
15'—
10'—
M99
5'—
0'—
N

M99 stands out more than M98. The 5th magnitude star *6 Comae Berenices* serves as a marker for identifying both of these galaxies.

For easy comparison, all photographs are shown at the same scale, measuring 2.3° x 1.3°.

M 100 The Mirror ✧ Virgo Cluster Galaxy

Original Messier Description

from the 1784 edition of *Connaissance des Temps*

Observed 1781. April 13

Nebula without stars, with the same brightness as the preceding one, located in the ear of Virgo. Observed by Mr. Méchain on March 15, 1781. The three nebulae, *Nos. 98, 99 & 100*, are very difficult to recognize because of the faintness of their light: you can only see them in good weather & near their meridian passage.

NGC Summary by J.L.E. Dreyer, circa 1888
NGC Number: 4321
Mag: 9.4   Size: 6.9'
NGC Description: Very remarkable!! Pretty faint, very large, round, very gradually then pretty suddenly becomes brighter in the middle to a mottled nucleus, 2-branch spiral.

Location
Constellation: Coma Berenices (Com)

Year 2000 Coordinates
RA: 12h 22.9m
Dec: +15° 49'

Observation Periods
Evenings 8 p.m. : March to August
Mornings 4 a.m. : November to April

Facts
Name: The Mirror (of M99)
Type of Object: Spiral galaxy
Magnitude: 9.4
Distance: 41,000,000 ly
Physical Size: 83,000 ly in diameter
Arc Degree Size in Sky: 7' x 6'
Other: The shape of this galaxy is classified as Sc, which means that it has a round nucleus with loosely wound arms. It is a member of the Virgo Galaxy Cluster.

Description of M100 using a 4-inch refractor at 48x

One of the faintest galaxies in this Virgo Galaxy Cluster catalogued by Messier. It has a faint, starlike center matching the brightness of stars around it. Appears face on.

Locating Index: Challenging because it is not near any conspicuously bright stars. To locate this galaxy, use Chart U, starting at the star *Vindemiatrix* and then "star hop" until you find it.

Identifying/Observing Index: Challenging because of its faintness; however, it can just be seen in slightly light-polluted skies.

TO LOCATE
Refer to
Star Charts
**A,
L, T
& U**

Near M100 is the cigar-shaped galaxy NGC 4312.

For easy comparison, all photographs are shown at the same scale, measuring 2.3° x 1.3°.

M101 Pinwheel Galaxy ✧ Galaxy

Original Messier Description

from the 1784 edition of *Connaissance des Temps*

Observed 1781. March 27
By Mr. Méchain, that Mr. Messier hasn't seen yet.

Nebula without stars, very obscure & quite large, 6 to 7 minutes in diameter, between the left hand of Boötes & the tail of Ursa Major. It's hard to distinguish it while illuminating the wires.

Location

Constellation: Ursa Major (UMa)

Year 2000 Coordinates
RA: 14h 03.2m
Dec: +54° 21'

Observation Periods

Evenings 8 p.m. : March to September
Mornings 4 a.m. : November to May

NGC Summary by J.L.E. Dreyer, circa 1888

NGC Number: 5457
Mag: 7.7 **Size:** 26.9'
NGC Description: Pretty bright, very large, irregularly round, gradually and then very suddenly becomes much brighter in the middle to a small nucleus.

Facts

Name: Like M33, M101 is also known as the "Pinwheel Galaxy"
Type of Object: Spiral galaxy
Magnitude: 7.7
Distance: 27,000,000 ly
Physical Size: 212,000 ly in diameter
Arc Degree Size in Sky: 27' x 26'
Other: The shape of this galaxy is classified as Sc, which means that it has a round nucleus with loosely wound arms. It is a very large spiral galaxy, with more than twice the diameter of our Milky Way Galaxy.

Description of M101 using a 4-inch refractor at 48x

Large in comparison to other Messier galaxies but very faint because it is a face-on spiral, thus it has low surface brightness. In this regard, it's similar to M33. Overall, its appearance is a roundish smudge with a slightly extended and brighter center.

Locating Index: Fairly easy because it forms a point of an equilateral triangle with the handle stars *Alkaid* and *Mizar* that make up the Big Dipper.

Identifying/Observing Index: Challenging because of its extended faintness. It easily fades in light-polluted skies.

ARC SCALE

1° —
3/4° —
1/2° —
25' —
20' —
15' —
10' —
5' —
0' —

N

TO LOCATE
Refer to
Star Charts
A,
K
& **M**

None of the spiral structure seen in this picture can be detected with a small telescope. I personally have difficulty seeing this galaxy because overall, its surface brightness is very low.

For easy comparison, all photographs are shown at the same scale, measuring 2.3° x 1.3°.

M102 Méchain's Lost Galaxy ✦ Galaxy

Original Messier Description

from the 1784 edition of *Connaissance des Temps*

Observed 1781. March 27
By Mr. Méchain, that Mr. Messier hasn't seen yet.

Nebula between the stars θ of Bootes &
ι of Draco: it's very faint; near it there's
a star of sixth magnitude.

Author's Note: Many books list M102 as a duplication of M101,
however it is not clear from the above description whether this is the
case. This description of M102, which also indicates its location, fits
the description and location of the galaxy NGC 5866, which would
have been visible with scopes used by Messier and Méchain.
Méchain later looked for NGC 5866 but could not see it, so he reported M102 as a negative find or mistake. I think that it is quite possible
that Méchain originally saw and recorded NGC 5866 as M102, but
subsequently was unable to see it, probably because of telescope
limitations and/or marginal to poor seeing conditions. I urge you to
treat M102 as NGC 5866 and not as a duplicate of M101, so you can
better understand the observing challenges and telescope limitations
of the era. There is an excellent discussion of the M101/102 contro-
versary at the website www.seds.org.

NGC Summary   by J.L.E. Dreyer, circa 1888

NGC Number: 5866
Mag: 10.0   **Size:** 5.2'
NGC Description: Very bright, considerably large, pretty much extended
along position angle 146°, gradually
becomes brighter in the middle.

Location

Constellation: Draco (Dra)
Year 2000 Coordinates
RA: 15h 06.5m
Dec: +55° 46'

Observation Periods

Evenings 8 p.m. : April to October
Mornings 4 a.m. : December to June

Facts

Name: Méchain's Lost Galaxy
Type of Object: Elliptical galaxy
Magnitude: 9.9
Distance: 38,000,000 ly
Physical Size: Spans at least 66,000 ly
Arc Degree Size in Sky: 6' x 3'
Other: This elliptical galaxy is a Seyfert
galaxy and is classified as E6p. Seyfert
galaxies have very small but bright nuclei
with strange spectra, possibly the result
of active supermassive black holes. E6p
refers to a very elongated elliptical galaxy
that has peculiarities in its shape.

Description of M102 using a 4-inch refractor at 48x

Fainter and small. It appears to be more edge on, and has a brighter, starlike center with fainter ends on both sides. More of the galaxy is apparent with averted vision. To me, it's easier to see than M101. Well within the telescope capability of Messier and Méchain.

Locating Index: Challenging because it is not near any conspicuously bright stars. Try pointing the telescope about 1° south of the 5th magnitude star shown "above" M102 on Chart M.

Identifying/Observing Index: Slightly challenging because it is not very large or very bright. It's easily glimpsed in dark skies but fades with light pollution.

ARC SCALE

1°—
3/4°—
1/2°—
25'—
20'—
15'—
10'—
5'—
0'—

• 6

TO LOCATE
Refer to
Star Charts
A,
K
& **M**

Méchain's reference to a 6th magnitude star near M102, centered, is probably the 6½ magnitude star indicated by the number 6.

For easy comparison, all photographs are shown at the same scale, measuring 2.3° x 1.3°.

M103 ✧ Open Cluster

Original Messier Description

from the 1784 edition of *Connaissance des Temps*

Observed 1781. March 27
By Mr. Méchain, that Mr. Messier hasn't seen yet.

Cluster of stars between ε & δ in the leg of Cassiopeia.

Author's Notes: M103 is the last object listed for the final publication of the 1784 catalogue in *Connaissance des Temps*.
 Objects M104 through M110 were added later, in the 1900s, based on notes, correspondence or drawings Messier and Méchain made about these objects while exploring the sky.

Location

Constellation: Cassiopeia (Cas)

Year 2000 Coordinates
RA: 1h 33.2m
Dec: +60° 42'

Observation Periods

Evenings 8 p.m. : September to March
Mornings 4 a.m. : May to November

Facts

Name: No common name
Type of Object: Open cluster
Magnitude: 7
Distance: 8,500 ly
Physical Size: Spans 15 ly
Arc Degree Size in Sky: Extends 6'
Other: This cluster contains about 25 stars with the brightest shining at magnitude 10.6. Its age is estimated at 22,000,000 years.

NGC Summary by J.L.E. Dreyer, circa 1888

NGC Number: 581
Mag: 7.4 **Size:** 6'
NGC Description: Cluster, pretty large, bright, round, rich in stars, contains stars of magnitude 10 to 11.

Description of M103 using a 4-inch refractor at 48x

Shaped like an arrow. In a very thick area of the Milky Way Band, this cluster blends in with the surrounding stars. Viewing with averted vision makes it more pronounced. There are many clumps of stars in this area.

Locating Index: Easy because it is 1° away from δ *Cassiopeiae*.

Identifying/Observing Index: Slightly challenging because it does not "pop" as much as you might think. In darker skies, it's surrounded by many similar magnitude stars that make it harder to distinguish this cluster from the "background." Because of this, I sometimes have difficulty seeing it. Easier in light-polluted skies.

TO LOCATE
Refer to
Star Charts
A,
C
& **E**

M103, centered, is easy to find because of its location near the bright 3rd magnitude star *Ruchbah*, but in dark skies, it blends in with the other Milky Way stars and is not as apparent or distinct as it appears in this picture.

For easy comparison, all photographs are shown at the same scale, measuring 2.3° x 1.3°.

M104 Sombrero Galaxy ✧ Galaxy

Original Descriptions*

Messier's description of M104 comes from a handwritten note in his 1784 copy of *Connaissance des Temps*. It reads:

Very faint nebula, seen by M. Méchain on May 11, 1781.

Méchain, in his May 6, 1783 letter to Bernoulli describes M104 as follows:

On May 11, 1781, I discovered a nebula above the Raven which did not appear to me to contain a single star. It is of a faint light and difficult to find if the micrometer wires are illuminated. I have compared it on this day and the following with Spica in the Virgin and from this derived its right ascension 187^d 9' 42" and its southern declination 10^d 24' 49". It does not appear in the *Connaissance des Temps*.

NGC Summary by J.L.E. Dreyer, circa 1888

NGC Number: 4594
Mag: 8.3   **Size:** 8.9'
NGC Description: Remarkable! Very bright, very large, extremely extended along position angle 92°, very suddenly becomes much brighter in the middle to the nucleus.

Location

Constellation: Virgo (Vir)
Year 2000 Coordinates
RA: 12h 40.0m
Dec: −11° 37'

Observation Periods

Evenings 8 p.m. : April to July
Mornings 4 a.m. : December to March

Facts

Name: Sombrero Galaxy
Type of Object: Spiral galaxy
Magnitude: 8.3
Distance: 48,000,000 ly
Physical Size: 126,000 ly in diameter
Arc Degree Size in Sky: 9' x 4'
Other: The shape of this galaxy is classified as Sb, which means that it has a round nucleus with medium wound arms; however, it has a huge bulging nucleus and an exceptionally dark dust lane around its rim that explains its name.

*Translated text from www.seds.org website.

Description of M104 using a 4-inch refractor at 48x

Near three 8th magnitude stars that almost point to the galaxy. Visually, this galaxy looks unique. The core appears extended on both sides of a bright/dark line that evenly divides the galaxy. The bright/dark line is best glimpsed with averted vision.

Locating Index: Challenging because it is not near any conspicuously bright stars. It almost forms the right angle of a triangle with the stars *Spica* and *δ Corvi*.

Identifying/Observing Index: Fairly easy, especially in dark skies. Its brightness is on the order of the brightest in the Virgo Galaxy Cluster. It will be more difficult to see in light-polluted skies.

TO LOCATE Refer to Star Charts **A, L** & **N**

ARC SCALE

1° —
3/4° —
1/2° —
25' —
20' —
15' —
10' —
5' —
0' —

The three 8th magnitude stars that point to the Sombrero are less than 30' away.

For easy comparison, all photographs are shown at the same scale, measuring 2.3° x 1.3°.

M 105 ✧ Galaxy

Original Méchain Description*

Méchain, in his May 6, 1783 letter to Bernoulli, describes M105 as follows:

Mr. Messier mentions there on page 264 and 265 two nebulous stars, which I have discovered in the Lion. I find nothing to correct for the given positions which I have determined by comparison of their situation with respect to Regulus. There is, however, a third one, somewhat more northerly, which is even more vivid than the two preceding ones. I discovered this one on March 24, 1781, 4 or 5 days after I had found the other two. On April 10, I compared its situation with Gamma Leonis from which followed its right ascension 159^d 3' 45" and its northern declination of 13^d 43' 58".

Location

Constellation: Leo (Leo)

Year 2000 Coordinates
RA: 10h 47.8m
Dec: +12° 35'

Observation Periods

Evenings 8 p.m. : February to July
Mornings 4 a.m. : October to March

Facts

Name: No common name
Type of Object: Elliptical galaxy
Magnitude: 9.3
Distance: 22,000,000 ly
Physical Size: Spans at least 40,000 ly
Arc Degree Size in Sky: 5' x 4'
Other: The shape of this galaxy is classified as an E1, which means that it resembles a slightly elongated sphere.

*Translated text from www.seds.org website.

Description of M105 using a 4-inch refractor at 48x

M105 is the faintest of the M95, M96 and M105 trio. It is a little farther from M96 than M96 is from M95. Its center is star-like and if you look directly at it, it looks like a faint star. Use averted vision to see more of this elliptical.

Locating Index: Fairly easy because it forms the apex of a shallow isosceles triangle with the bright stars *Regulus* and *Chertan*.

Identifying/Observing Index: Fairly easy because the three galaxies, M95, M96 and M105 are near one another and are relatively bright. M105 is the faintest and M96 is the brightest.

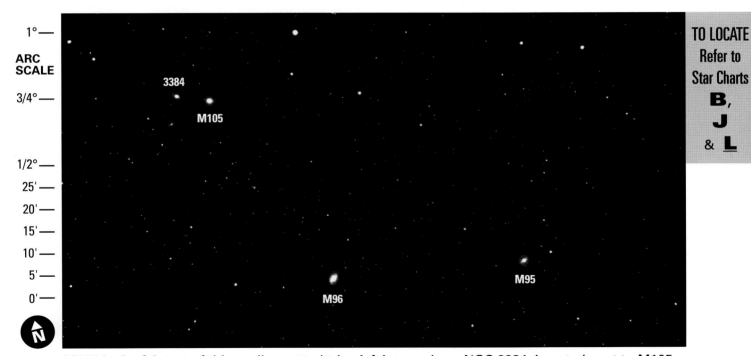

TO LOCATE
Refer to
Star Charts
B,
J
& **L**

M105 is the faintest of this easily spotted trio. A fainter galaxy, NGC 3384, is noted next to M105. Can you see it through your telescope?

For easy comparison, all photographs are shown at the same scale, measuring 2.3° x 1.3°.

M106 ✧ Galaxy

Original Méchain Description*

Méchain, in his May 6, 1783 letter to Bernoulli, describes M106 as follows:

In July 1781 I found another nebula close to the Great Bear near the star No. 3 of the Hunting Dogs and 1 deg more south, I estimate its right ascension 181^d 40' and its northern declination about 49^d. I am going to determine the more accurate position of this one shortly.

Location

Constellation: Canes Venatici (CVn)

Year 2000 Coordinates
RA: 12h 19.0m
Dec: +47° 18'

Observation Periods

Evenings 8 p.m. : February to August
Mornings 4 a.m. : October to April

Facts

Name: No common name
Type of Object: Spiral galaxy
Magnitude: 8.3
Distance: 26,000,000 ly
Physical Size: 136,000 ly in diameter
Arc Degree Size in Sky: 18' x 8'
Other: The shape of this galaxy is classified as Sb, which means that it has a round nucleus with medium wound arms.

NGC Summary by J.L.E. Dreyer, circa 1888

NGC Number: 4258
Mag: 8.3 **Size:** 18.2'
NGC Description: Very bright, very large, very much extended along position angle 0°, suddenly becomes brighter in the middle to the bright nucleus.

*Translated text from www.seds.org website.

Description of M106 using a 4-inch refractor at 48x

Pretty impressive galaxy because of its size. I find it to be very similar to M63.

Locating Index: Challenging because it is not near any conspicuously bright stars. It is "about" halfway between the star *Phad* of the Big Dipper and *α Canum Venaticorum*.

Identifying/Observing Index: Fairly easy in dark skies but it will be more challenging in light-polluted skies.

4217

TO LOCATE
Refer to
Star Charts
A/B,
K
& **M**

M106 is a fairly large galaxy; however, it is harder to find than most of the other Messier objects. Note the interesting set of stars to the right, housing the fainter galaxy NGC 4217.

For easy comparison, all photographs are shown at the same scale, measuring 2.3° x 1.3°.

319

M107 ✧ Globular Cluster

Original Méchain Description*

Méchain, in his May 6, 1783 letter to Bernoulli, describes M107 as follows:

In April 1782 I discovered a small nebula in the left flank of Ophiuchus between the stars Zeta and Phi, the position of which I have not yet observed any more closely.

Location

Constellation: Ophiuchus (Oph)

Year 2000 Coordinates
RA: 16h 32.5m
Dec: −13° 03'

Observation Periods

Evenings 8 p.m. : June to September
Mornings 4 a.m. : February to May

Facts

Name: No common name
Type of Object: Globular cluster
Magnitude: 8.1
Distance: 21,000 ly
Physical Size: 61 ly in diameter
Arc Degree Size in Sky: 10' in diameter
Other: The compactness of the stars in this globular cluster is very loose compared to the average.

NGC Summary by J.L.E. Dreyer, circa 1888

NGC Number: 6171
Mag: 8.1   **Size:** 10'
NGC Description: Globular cluster of stars, large, very rich in stars, very much compressed, round, well resolved.

*Translated text from www.seds.org website.

Description of M107 using a 4-inch refractor at 48x

Very diffused and faint, reminding me more of a galaxy. A bit of a center is visible but overall, the brightness of this globular is pretty even. One side appears more extended than the other.

Locating Index: Fairly easy because it forms the right angle in a narrow or skinny right triangle with *Sabik* and ζ *Ophiuchi*.

Identifying/Observing Index: Challenging because it is diffused and faint. It will be difficult to see in light-polluted skies.

TO LOCATE Refer to Star Charts **A, N** & **P**

The reason that Messier probably missed seeing M107 (centered) earlier in his cataloguing is that it is much fainter than other globulars in the area.

For easy comparison, all photographs are shown at the same scale, measuring 2.3° x 1.3°.

321

M 108 ✧ Galaxy

Original Descriptions*

Messier makes note of M108 in his description of M97. Also, in a manuscript, Messier describes M108: Nebula near the preceding — it is even fainter: it is 48 or 49' farther north and 30 min — following in RA: Found by M. Méchain 2 or 3 days after the preceding [M97, which he found on February 16, so M108 was found on February 18 or 19, 1781].

Méchain in his May 6, 1783 letter to Bernoulli describes M108 as follows: Page 265 No. 97 [M97]. A nebula near Beta in the Great Bear. Mr. Messier mentions, when indicating its position, two others, which I also have discovered and of which one is close to this one [M108], the other [M109] is situated close to Gamma in the Great Bear, but I could not yet determine their positions.

Author's Note: Bracketed text is not original, but added to clarify meaning.

NGC Summary by J.L.E. Dreyer, circa 1888

NGC Number: 3556
Mag: 10.1   **Size:** 8.3'
NGC Description: Considerably bright, very large, very much extended along position angle 79°, pretty brighter in the middle, not resolved.

Location

Constellation: Ursa Major (UMa)

Year 2000 Coordinates
RA: 11h 11.5m
Dec: +55° 40'

Observation Periods

Evenings 8 p.m. : February to August
Mornings 4 a.m. : October to April

Facts

Name: No common name
Type of Object: Spiral galaxy
Magnitude: 10.0
Distance: 24,000,000 ly
Physical Size: 56,000 ly in diameter
Arc Degree Size in Sky: 8' x 2'
Other: The shape of this galaxy is classified as Sc, which means that it has a round nucleus with loosely wound arms.

*Translated text from www.seds.org website.

Description of M108 using a 4-inch refractor at 48x

Appears much fainter, about half the brightness of the nearby Owl Nebula, M97. Shape is more edge on, resembling a cigar, like M82.

Locating Index: Easy because it is about 1½° away from the 2nd magnitude star *Merak* that makes up the bottom of the bowl of the Big Dipper. Additionally, M108 is on the same line as the bottom of the bowl.

Identifying/Observing Index: Challenging because it is very faint. You need dark skies because it fades in light pollution.

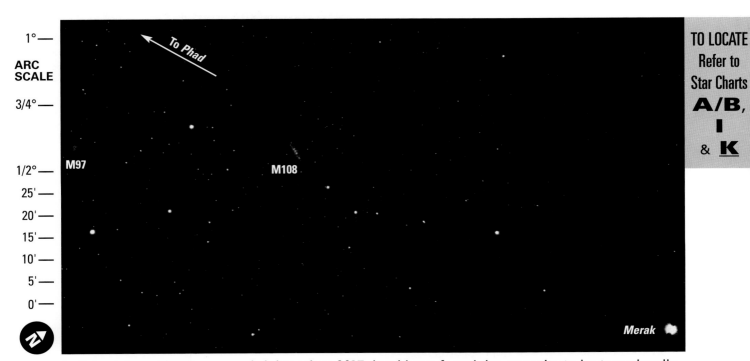

TO LOCATE
Refer to
Star Charts
A/B,
I
& **K**

In this picture, M108 appears brighter than M97, but I have found the opposite to be true visually.

For easy comparison, all photographs are shown at the same scale, measuring 2.3° x 1.3°.

M 109 ✦ Galaxy

Original Descriptions*

Messier makes note of M109 in his description of M97. Also, in a manuscript, Messier describes M109: Nebula near Gamma UMa, same right ascension a bit near this star and 1 deg — more south. Discovered by M. Méchain on March 12, 1781.

Méchain in his May 6, 1783 letter to Bernoulli describes M109 as follows: Page 265 No. 97 [M97]. A nebula near Beta in the Great Bear. Mr. Messier mentions, when indicating its position, two others, which I also have discovered and of which one is close to this one [M108], the other [M109] is situated close to Gamma in the Great Bear, but I could not yet determine their positions.

Author's Note: Bracketed text is not original, but added to clarify meaning.

NGC Summary by J.L.E. Dreyer, circa 1888

NGC Number: 3992
Mag: 9.8   **Size:** 7.6'
NGC Description: Considerably bright, very large, pretty much extended, suddenly becomes brighter in the middle to a bright mottled nucleus.

Location

Constellation: Ursa Major (UMa)

Year 2000 Coordinates
RA: 11h 57.6m
Dec: +53° 23'

Observation Periods

Evenings 8 p.m. : February to August
Mornings 4 a.m. : October to April

Facts

Name: No common name
Type of Object: Spiral galaxy
Magnitude: 9.8
Distance: 27,000,000 ly
Physical Size: 63,000 ly in diameter
Arc Degree Size in Sky: 8' x 5'
Other: The shape of this galaxy is classified as SBb, which means that it has an extended nucleus with a "bar" or "straight arm" that goes through it, and medium wound arms (see page 340).

*Translated text from www.seds.org website.

Description of M109 using a 4-inch refractor at 48x

Fainter than the nearby M108 next to *Merak*. This spiral galaxy is more face on, and ovalish, which probably accounts for its overall low brightness. About the same size as the Owl Nebula, M97. "Between" two 9th magnitude stars but closer to one than the other.

Locating Index: **Easy** because it is only about a Moon's width from the 2nd magnitude star *Phad* which makes up the bottom of the bowl in the Big Dipper.

Identifying/Observing Index: **Challenging** because it is very faint. It needs dark skies. and it will fade with any amount of light pollution.

TO LOCATE Refer to Star Charts **A/B**, **I** & **K**

M109 is easy to find because it is within a low-power eyepiece's field of view from the 2nd magnitude star *Phad* of the Big Dipper, but more difficult to see because it is so faint.

For easy comparison, all photographs are shown at the same scale, measuring 2.3° x 1.3°.

M 110 ✧ Galaxy

Original Messier Description*

A description of M110 by Messier appeared in the 1801 issue of *Connaissance des Temps*, **however, the observation was made 25 years earlier. It reads:** On August 10, [1773] I examined, under a very good sky, the beautiful nebula of the girdle of Andromeda, with my achromatic refractor, which I had set to magnify 68 times, to create a drawing like the one in Orion (Mém. de l'Acad. 1771, pag. 460). I saw that which C. le Gentil discovered on October 29, 1749 [M32]. I also saw a new, fainter one, located north of the great [nebula], which was distant from it about 35' in right ascension and 24' in declination. It appeared amazing to me that this faint nebula has escaped the astronomers and myself, since the discovery of the great [nebula] by Simon Marius in 1612, because when observing the great [nebula], the small one is located in the same field of the telescope. I will do a drawing of that remarkable nebula in the girdle of Andromeda, with the two small ones which accompany it.

Another description of M110 by Messier is noted in his drawing of M31, which was published in 1807 but originally drawn in 1773. The description says: Small nebula, very faint.

Author's Note: Bracketed text is not original, but added to clarify meaning.

NGC Summary by J.L.E. Dreyer, circa 1888

NGC Number: 205
Mag: 8.0 **Size:** 17.4'
NGC Description: Very bright, very large, much extended along position angle 165°, very gradually becomes very much brighter in the middle.

Location

Constellation: Andromeda (And)
Year 2000 Coordinates
RA: 0h 40.4m
Dec: +41° 41'

Observation Periods

Evenings 8 p.m. : August to March
Mornings 4 a.m. : April to November

Facts

Name: No common name
Type of Object: Elliptical galaxy
Magnitude: 8.0
Distance: 2,400,000 ly
Physical Size: Spans at least 12,000 ly
Arc Degree Size in Sky: 17' x 10'
Other: The shape of this galaxy is classified as E6, which means that it is extremely elongated. This galaxy is a companion to M31, the Andromeda Galaxy, because it is gravitationally bound to it like a moon. It is also a member of our Local Group of galaxies.

*Translated text from www.seds.org website.

Description of M110 using a 4-inch refractor at 48x

Much fainter, but larger and more diffused than M32. Use averted vision to make it "pop" more. Visually, this is a very good example of an elliptical galaxy because it is very large in arc size.

Locating Index: Fairly easy once you locate M31, the Andromeda Galaxy, because it is about ½°, or one Moon's diameter, from the center of the core. Look for it somewhat opposite the core from M32.

Identifying/Observing Index: Slightly challenging, because it fades in light-polluted skies. Look for something fainter and more diffused than M32.

TO LOCATE
Refer to
Star Charts
**B,
C**
& **E**

M110 is close to being opposite the core of M31 from M32. It is *much* fainter than M32.

For easy comparison, all photographs are shown at the same scale, measuring 2.3° x 1.3°.

327

M 111 Double Cluster ✧ Cluster, "Western Half"

HONORARY ENTRY

The Double Cluster is an anomalous omission in Messier's catalogue which can only be attributed to repeated oversights because evidence indicates that he knew about the side-by-side clusters.

The Double Cluster has been known since antiquity as a fuzzy patch in the sky because it can be seen with the naked eye. In Ptolemy's *Almagest* from circa A.D. 175, it is described as "The nebulous mass on the right hand" of Perseus. Messier drew charts with Perseus' hand at the exact location of the nebulous mass. Also, Messier makes reference, in his first published catalogue of 1772, to De Chéseaux's list of 21 objects, which includes the Double Cluster, giving each cluster a separate entry.

Messier's final published catalogue of 1784 listed 103 objects. Future versions of the catalogue, which would have included additional objects, were halted because William Herschel published a catalogue of 2,000 objects by 1789.

Continues on page 330 under the same heading

NGC Summary by J.L.E. Dreyer, circa 1888

NGC Number: 869
Mag: 4 **Size:** 30'
NGC Description: Remarkable! Cluster, *very* large, very rich in stars, contains stars of magnitude 7 to 14.

Location

Constellation: Perseus (Per)
Year 2000 Coordinates
RA: 02h 19.0m
Dec: +57° 09'

Observation Periods

Evenings 8 p.m. : September to March
Mornings 4 a.m. : June to November

Facts

Name: *"Western Half"* of the Double Cluster
Type of Object: Open cluster
Magnitude: Estimates range from 3.5 to 5.3
Distance: 7,100 ly
Physical Size: Spans 62 ly
Arc Degree Size in Sky: 30'
(Together, the pair extends across 1°)
Other: Contains about 200 stars with the brightest shining at magnitude 6.6. Its age is estimated at 5,600,000 years.

Description of M111 using a 4-inch refractor at 48x

Magnificent — two beautiful clusters, side by side, spread across 1° of sky! Either by itself would be beautiful, but together they form a natural celestial "wonder." The cluster that has the higher concentration of stars at its center is M111, or NGC 869.

Locating Index: Little challenging if you cannot see it with your naked eye, otherwise **easy** if you can. It lies about halfway on a line between γ *Persei* and δ *Cassiopeiae*.

Identifying/Observing Index: Easy because it is so big and bright. It's best observed with low power to capture both clusters in the same eyepiece field of view.

TO LOCATE Refer to Star Charts **B**, **C** & **E**

Arc scale markings: 1°, 3/4°, 1/2°, 25', 20', 15', 10', 5', 0'

ARC SCALE

M112 or NGC 884   M111 or NGC 869

N

The Double Cluster was a fuzzy patch in the sky known to the ancients, and it is a pity they never got to see it through a telescope. Look for the two red stars indicated by the arrows.

For easy comparison, all photographs are shown at the same scale, measuring 2.3° x 1.3°.

M 112 Double Cluster ✧ Cluster, "Eastern Half"

HONORARY ENTRY

Continued from the same section on page 328

Objects 104 to 110 were added in the 20th century from historical evidence indicating that Messier and Méchain had knowledge of these objects.

I have decided, in this book, to rectify Messier's biggest oversight and honor him by adding the Double Cluster as M111 and M112, thus "completing" his catalogue.

In 1801, Messier wrote in the *Connaissance des Temps* that he was planning to publish a future catalogue listing additional objects (which never was published). He added that his catalogue was different from Herschel's in that it represented a set of objects visible with smaller telescopes. The addition of M103 through M110 and now my addition of M111 and M112 are in keeping with his desires.

I have a more lengthy discussion about this topic starting on page 44.

I have a more lengthy discussion about this topic starting on page 44.

NGC Summary by J.L.E. Dreyer, circa 1888
NGC Number: 884
Mag: 4 **Size:** 30'
NGC Description: Remarkable! Cluster, very large, very rich in stars, ruby star in the middle.

Location
Constellation: Perseus (Per)
Year 2000 Coordinates
RA: 02h 22.4m
Dec: +57° 07'

Observation Periods
Evenings 8 p.m. : September to March
Mornings 4 a.m. : June to November

Facts
Name: *"Eastern Half"* of the Double Cluster
Type of Object: Open cluster
Magnitude: Estimates range from 3.6 to 6.1
Distance: 7,500 ly
Physical Size: Spans 65 ly
Arc Degree Size in Sky: 30' (Together, the pair extends across 1°.)
Other: Contains about 115 stars with the brightest shining at magnitude 8.1. Its age is estimated at 3,200,000 years.

Description of M112 using a 4-inch refractor at 48x

It boasts a red star near its "core." This is the cluster whose stars appear to have a looser arrangement *and* a less condensed "center" than its other half. In my Tucson skies, I can often see both clusters as a fuzzy patch with my naked eyes.

Locating Index: Little challenging if you cannot see it with your naked eye, otherwise **easy** if you can. It lies about halfway on a line between γ *Persei* and δ *Cassiopeiae*.

Identifying/Observing Index: Easy because it is so big and bright. It's best observed with low power to capture both clusters in the same eyepiece field of view.

TO LOCATE Refer to Star Charts **B, C** & **E**

ARC SCALE

1° —
3/4° —
1/2° —
25' —
20' —
15' —
10' —
5' —
0' —

N

M112 or NGC 884   M111 or NGC 869

Some modern star charts designate NGC 884 with χ and NGC 869 with *h*. However, Stephen O'Meara and Daniel Green believe that "χ"was the original designation for the combined Double Cluster and that "*h*" indicated the two stars as shown.   For easy comparison, all photographs are shown at the same scale, measuring 2.3° x 1.3°.   331

CATALOGUE
OF THE NEBULAE
AND
STAR CLUSTERS

that can be seen amongst the fixed stars,
on the horizon in Paris.

Observed at the Navy observatory,
with various instruments.

By Mr. Messier.

February 16, 1771

Several astronomers have searched for nebular stars, including Hévélius, Huygens, Derham, Halley, Cheseaux, Father De la Caille and, finally, Mr. le Gentil; other astronomers have discovered some by chance, either while working on establishing the location of stars for cataloguing purposes, or while observing comets; several of these astronomers only indicated the constellations where they were located, without giving their exact position or a detailed description.

I started this book in 1764, by observing those which were already known as well as researching others that had eluded astronomers since the telescope was invented; this long and tedious work allows me today to submit to the Académie a more comprehensive, more precise and more detailed catalogue of nebular stars, a reference work that may have been lacking in the field of astronomy.

The 1758 comet, on August 28th, was between the horns of *Taurus*. I discovered above the southern horn and at a short distance from the ζ star of this constellation, a whitish, elongated light in the shape of a candle's light, that did not contain any stars. This light was about the same as those of the comet I had been observing at the time; however it was a little brighter, whiter and a little more elongated than the comet's, which had always appeared to have an almost round coma, with no apparent tail nor beard. On September 12th of that year, I established the position of this nebula, its right ascension being measured at 80° 0' 33" and its declination at 21° 45' 27" north. This nebula was placed on the map of the 1758 comet's apparent path.

On September 11, 1760, I discovered in the head of Aquarius a beautiful nebula that does not contain any stars; I studied it with a good, 30-inch lens Gregorian telescope that magnified one hundred and four times; its center is bright, the surrounding nebulosity is round; it bears a fair resemblance to the nebula found between Sagittarius' head and bow: it may have a 4-minute large circle diameter; it is very visible with a regular 2-foot telescope; I compared its meridian passage with that of α Aquarii, which was located on the same parallel; its right ascension was established at 320° 17' and its declination at 1° 47' south. On the night of July 26–27, 1764, I saw this nebula for a second time; it was the same, with the same appearance. This nebula can be found on the map of the famous Halley comet, which I observed when it returned in 1759.

Above left. The first page of the first edition of Messier's catalogue as it appears in the 1772 *Mémoires de l'Académie des Sciences. Above.* Translations of the first two pages of the catalogue, from a total of 24 pages, 19½ pages of discussion and 4½ catalogue pages. *Next page.* First two pages listing objects from the first catalogue. Note that the objects are not numbered and the descriptions are shorter than what appeared in the third and final edition of the catalogue published in the 1784 edition of *Connaisssance des Temps* (see page 105 in this book). Images courtesy of the Institut de France Académie des Sciences Archives.

Appendices

de 19ᵈ 24' 20" d'ascension droite, & 31ᵈ 29' 0" de déclinaison australe.

Le 29 Juillet 1764, le ciel étoit serein; mais il y avoit un peu de vapeurs à l'horizon, j'ai cherché cette nébuleuse sans pouvoir la découvrir.

M. Cassini rapporte dans ses Élémens d'Astronomie, *page 79*, que son père découvrit une nébuleuse dans l'espace qui est entre le grand Chien & le petit Chien, qui est une des plus belles qu'on voie par la lunette.

J'ai cherché plusieurs fois cette nébuleuse, par un ciel serein, sans pouvoir la découvrir; il y a lieu de présumer que c'étoit une Comète qui commençoit ou qui cessoit de paroître, puisque rien ne ressemble tant à une étoile nébuleuse, qu'une Comète qui commence à être visible aux instrumens.

TABLE des Nébuleuses, ainsi que des amas d'Étoiles, que l'on découvre parmi les Étoiles fixes sur l'horizon de Paris; observées à l'Observatoire de la Marine.

ANNÉES & JOURS	ASCENSION droite	DÉCLINAISON	DIAM.	INDICATION DES NÉBULEUSES & amas d'Étoiles.
	D. M. S.	D. M. S.	D. M.	
1758. Sept. 12	80. 0. 33	21. 45. 27. B.	....	nébuleuse placée au-dessus de la corne méridionale du Taureau.
1760. Sept. 11	320. 17. 0	1. 47. 0. A.	0. 4	nébuleuse sans étoile, dans la tête du Verseau.
1764. Mai... 3	202. 51. 19	29. 32. 57. B.	0. 3	nébuleuse sans étoile, entre la queue & les pattes d'un des Chiens de chasse d'Hévélius.
8	242. 16. 56	23. 55. 40. A.	0. 2½	amas de très-petites étoiles, près d'Antarès & sur son parallèle.
23	226. 39. 4	2. 57. 16. B.	0. 3	belle nébuleuse sans étoile, entre le Serpent & la Balance, près de l'étoile de 6.ᵉ grandeur, cinquième du Serpent; suivant le catalogue de Flamsteed.

ANNÉE & JOURS	ASCENSION droite	DÉCLINAISON	DIAM.	INDICATION DES NÉBULEUSES & amas d'Étoiles.
	D. M. S.	D. M. S.	D. M.	
1764. Mai. 23	261. 10. 39	32. 10. 34. A.	0. 15	amas de petites étoiles entre l'arc du Sagittaire & la queue du Scorpion.
23	264. 30. 24	34. 40. 34. A.	0. 30	amas d'étoiles, peu éloigné du précédent, entre l'arc du Sagittaire & la queue du Scorpion.
23	267. 29. 30	24. 21. 10. A.	0. 30	amas d'étoiles entre l'arc du Sagittaire & le pied droit d'Ophiucus cet amas contient l'étoile de 7.ᵉ grandeur, la 9.ᵉ du Sagittaire, suivant le catalogue de Flamsteed.
28	256. 20. 36	18. 13. 26. A.	0. 3	nébuleuse sans étoile, dans la jambe droite d'Ophiucus, entre les étoiles η & ρ de cette constellation.
29	251. 12. 6	3. 42. 18. A.	0. 4	nébuleuse sans étoile, dans la ceinture d'Ophiucus, près de la 30.ᵉ étoile de cette constellation, suivant Flamsteed.
30	279. 35. 43	6. 31. 1. A.	0. 4	amas d'un grand nombre de petites étoiles, près de l'étoile k d'Antinoüs.
30	248. 43. 10	1. 30. 28. A.	0. 3	nébuleuse sans étoile, dans le Serpent, entre le bras & le côté gauche d'Ophiucus.
Juin 1	248. 18. 48	36. 54. 44. B.	0. 3	nébuleuse sans étoile, dans la ceinture d'Hercule, à 2 degrés au-dessous de l'étoile η de cette constellation.
1	261. 18. 29	3. 5. 45. A.	0. 2	nébuleuse sans étoile dans la draperie qui passe par le bras droit d'Ophiucus, sur le parallèle de l'étoile ζ du Serpent.
3	319. 40. 19	10. 40. 3. B.	0. 3	nébuleuse sans étoile, entre les têtes de Pégase & du petit Cheval.
3	271. 15. 3	13. 51. 44. A.	0. 8	amas de petites étoiles mêlées de nébulosités, proche la queue du Serpent, à peu de distance du parallèle de l'étoile ξ de cette constellation.

The 88 Constellations

Constellation Name	Abbreviation	Meaning of Name	Genitive Form*
ANDROMEDA	And	Daughter of Cassiopeia	*Andromedae*
ANTLIA	Ant	Air Pump	*Antliae*
APUS	Aps	Bird of Paradise	*Apodis*
AQUARIUS	Aqr	Water Bearer	*Aquarii*
AQUILA	Aql	Eagle	*Aquilae*
ARA	Ara	Altar	*Arae*
ARIES	Ari	Ram	*Arietis*
AURIGA	Aur	Charioteer	*Aurigae*
BOOTES or BOÖTES	Boo	Herdsman	*Bootis*
CAELUM	Cae	Engraving Tool	*Caeli*
CAMELOPARDALIS	Cam	Giraffe	*Camelopardalis*
CANCER	Cnc	Crab	*Cancri*
CANES VENATICI	CVn	Hunting Dogs	*Canum Venaticorum*
CANIS MAJOR	CMa	Big Dog	*Canis Majoris*
CANIS MINOR	CMi	Little Dog	*Canis Minoris*
CAPRICORNUS	Cap	Sea Goat	*Capricorni*
CARINA	Car	Ship's Keel	*Carinae*
CASSIOPEIA	Cas	Queen Cassiopeia	*Cassiopeiae*
CENTAURUS	Cen	Centaur	*Centauri*
CEPHEUS	Cep	King Cepheus	*Cephei*
CETUS	Cet	Whale	*Ceti*
CHAMAELEON	Cha	Chameleon	*Chamaeleontis*
CIRCINUS	Cir	Drawing Compass	*Circini*
COLUMBA	Col	Dove	*Columbae*
COMA BERENICES	Com	Berenice's Hair	*Comae Berenices*
CORONA AUSTRALIS	CrA	Southern Crown	*Coronae Australis*
CORONA BOREALIS	CrB	Northern Crown	*Coronae Borealis*
CORVUS	Crv	Crow	*Corvi*

*See bottom of page 336.

Constellation Name	Abbreviation	Meaning of Name	Genitive Form*
CRATER	Crt	Cup	*Crateris*
CRUX	Cru	Southern Cross	*Crucis*
CYGNUS	Cyg	Swan	*Cygni*
DELPHINUS	Del	Dolphin	*Delphini*
DORADO	Dor	Goldfish	*Doradus*
DRACO	Dra	Dragon	*Draconis*
EQUULEUS	Equ	Little Horse	*Equulei*
ERIDANUS	Eri	River Eridanus	*Eridani*
FORNAX	For	Furnace	*Fornacis*
GEMINI	Gem	Twins	*Geminorum*
GRUS	Gru	Crane	*Gruis*
HERCULES	Her	Son of Zeus	*Herculis*
HOROLOGIUM	Hor	Clock	*Horologii*
HYDRA	Hya	Sea Serpent	*Hydrae*
HYDRUS	Hyi	Water Snake	*Hydri*
INDUS	Ind	Indian	*Indi*
LACERTA	Lac	Lizard	*Lacertae*
LEO	Leo	Lion	*Leonis*
LEO MINOR	LMi	Little Lion	*Leonis Minoris*
LEPUS	Lep	Hare	*Leporis*
LIBRA	Lib	Scales	*Librae*
LUPUS	Lup	Wolf	*Lupi*
LYNX	Lyn	Lynx	*Lyncis*
LYRA	Lyr	Lyre	*Lyrae*
MENSA	Men	Table Mountain	*Mensae*
MICROSCOPIUM	Mic	Microscope	*Microscopii*
MONOCEROS	Mon	Unicorn	*Monocerotis*
MUSCA	Mus	Fly	*Muscae*
NORMA	Nor	Level	*Normae*
OCTANS	Oct	Octant	*Octantis*
OPHIUCHUS	Oph	Snake Holder	*Ophiuchi*
ORION	Ori	Hunter	*Orionis*

*See bottom of page 336.

Constellation Name	Abbreviation	Meaning of Name	Genitive Form*
PAVO	Pav	Peacock	*Pavonis*
PEGASUS	Peg	Winged Horse	*Pegasi*
PERSEUS	Per	Rescuer of Andromeda	*Persei*
PHOENIX	Phe	Phoenix	*Phoenicis*
PICTOR	Pic	Easel	*Pictoris*
PISCES	Psc	Fishes	*Piscium*
PISCIS AUSTRINUS	PsA	Southern Fish	*Piscis Austrini*
PUPPIS	Pup	Ship's Stern	*Puppis*
PYXIS	Pyx	Ship's Compass	*Pyxidis*
RETICULUM	Ret	Eyepiece Reticle	*Reticuli*
SAGITTA	Sge	Arrow	*Sagittae*
SAGITTARIUS	Sgr	Archer	*Sagittarii*
SCORPIUS	Sco	Scorpion	*Scorpii*
SCULPTOR	Scl	Sculptor's Apparatus	*Sculptoris*
SCUTUM	Sct	Shield	*Scuti*
SERPENS	Ser	Snake	*Serpentis*
SEXTANS	Sex	Sextant	*Sextantis*
TAURUS	Tau	Bull	*Tauri*
TELESCOPIUM	Tel	Telescope	*Telescopii*
TRIANGULUM	Tri	Triangle	*Trianguli*
TRIANGULUM AUSTRALE	TrA	Southern Triangle	*Trianguli Australis*
TUCANA	Tuc	Toucan	*Tucanae*
URSA MAJOR	UMa	Big Bear	*Ursae Majoris*
URSA MINOR	UMi	Little Bear	*Ursae Minoris*
VELA	Vel	Sail	*Velorum*
VIRGO	Vir	Virgin	*Virginis*
VOLANS	Vol	Flying Fish	*Volantis*
VULPECULA	Vul	Little Fox	*Vulpeculae*

* The Genitive Form of the constellations are used in conjunction with Bayer letters or Flamsteed numbers to name stars. For example, *Polaris*, in Ursa Minor, has a Bayer letter designation of α, so it can also be referred to as, using its genitive, α *Ursae Minoris* (pronounced "alpha Ursae Minoris"). Likewise, *Denebola* in Leo (see Chart T) can be written as β *Leonis* or *94 Leonis*. This system of naming is especially useful for stars that do not have proper names. Genitives are a possessive form leftover from Latin. Also see pages 80, 338, 339 & 341 for more about Bayer letters, Flamsteed numbers and Genitives.

Expanded Glossary

> **Words in *italics* are defined in this glossary.**

Abbreviations. Some common abbreviations used in astronomy are listed on page 352, the last page of this book.

Altazimuth Telescope Mount. A mount that moves in *altitude* and *azimuth*, allowing a telescope to move "up" and "down" vertically (altitude) and rotate horizontally to any compass point (azimuth). Everyone is familiar with this mount because it is the type used with binoculars at tourist attractions. This mount has become increasingly popular because of its simplicity and lower cost. Many amateur and modern professional telescopes use computer controlled altazimuth mounts. Often abbreviated as alt-az or altaz.

Altitude. For an *altazimuth telescope mount*, altitude refers to the movement of the telescope "up" and "down" vertically from the horizon to directly overhead. Altitude also refers to a measurement system, where the height of an object above the horizon is expressed in arc degrees. This measurement system ranges from 0° at the horizon to 90° at the *zenith* (of the observer). The same object in the sky can have different altitudes for observers at different locations on Earth.

Apochromatic (APO). Optical term that refers to the highest quality telescope optics which are free of spherical, chromatic and other optical aberrations. Usually associated with refractor telescopes.

Arc Degree (°). Unit of angular measurement used in astronomy. One arc degree is the same as one compass point degree. There are 360 arc degrees in a circle; each arc degree is divided into 60 *arc minutes*; and each *arc minute* is divided into 60 *arc seconds*. *Arc seconds* are further subdivided into tenths. The word "arc" is usually omitted when using this measurement system, which can cause confu-sion when minutes (referring to clock time) are used in the same dialogue as arc minutes or seconds. The Sun and Moon are both about ½ of an arc degree (30 arc minutes or 1,800 arc seconds) in angular diameter. Notation example: 6° 26' 3.2"

Arc Minute ('). 1/60 of an arc degree. The Moon is about 30 arc minutes in diameter. See *Arc Degree*.

Arc Second ("). 1/3,600 of an arc degree or 1/60 of an arc minute. See *Arc Degree*.

Asterism. A recognizable or distinguished group of stars. Sometimes a subgroup of a *constellation*. The Big Dipper is technically an asterism of the *constellation* Ursa Major. Many asterisms are much smaller and can only be seen with binoculars or a telescope.

Averted Vision. Using peripheral vision to view a faint object instead of looking directly at it. The use of averted vision is an observing technique that helps observers see faint objects that would otherwise go unnoticed. Averted vision can also be combined with slow movements of the telescope to further enhance detection of faint objects. Very faint objects slightly below the threshold of averted vision can sometimes be glimpsed with peripheral vision if the object **moves** through the peripheral field of view. This is accomplished by moving the telescope (pushing or with motor controls) back and forth over an area known to contain a faint deep sky object.

Azimuth. For an *altazimuth telescope mount*, azimuth refers to the horizontal movement of the telescope — the rotation of the telescope in a horizontal circle, to any compass point around the horizon. Azimuth is also part of a measurement system. Azimuth starts with 0° at true North and arcs eastwardly, through 360° of the compass.

Barred Spiral Galaxy. See *Galaxies*.

Bayer Letters. The formal name of the system of lowercase *Greek letters* assigned to the brightest stars in each of the *constellations*. These letters were initially assigned by Johann Bayer (1572–1625), a German astronomer, in his star atlas "Uranometria." See *Genitive*. Also see *Flamsteed Numbers* and page 80.

CCD (Charged-Coupled Device). A term that refers to a component in astronomical instrumentation that records images digitally. Amateurs use this term loosely to indicate a specialized astronomical camera that is similar to consumer digital cameras.

Celestial Equator. A *great circle* that is the projection of the Earth's equator onto the sky. The *celestial equator* has *declination* 0° (corresponding to a latitude of 0°).

Celestial Horizon. The meeting of the sky and Earth. At sea, the celestial horizon is where the sky and water meet, so it is unobstructed and perfectly round.

Celestial Meridian. A *great circle* that divides the sky into eastern and western halves. This circle passes through the observer's *zenith* and the North and South Celestial Poles. At any particular location on Earth, celestial objects are at their highest in the sky when they are on the meridian (unless they are below either Celestial Pole).

Celestial Sphere. At one time, it was thought that the Sun, Moon, Planets, comets and stars resided on the inside of a giant sphere called the celestial sphere. Today, this is a convenient term to indicate the visible globe of stars that surrounds us. In essence, it is the "canvas" on which the Universe is drawn.

The list of constellations during Messier's time was a little different than it is today. For example, Navis or Argo the Ship (see descriptions for M46, M47 and M93) has been replaced with Vela, Puppis and Carina. And at one point, Lalande drew on a chart the constellation "Custos Messium," replacing Cassiopeia, to honor his friend.

Constellation. A group of visible stars that has been assigned a name (most often by the Greeks). The stars in a constellation usually form a pattern that aids in their recognition. The visible stars were first categorized into constellations thousands of years ago and are associated with lore. Today the constellations are not just named groups of stars, but also include a specific, bounded area of the sky around the stars. There is a total of 88 constellations.

Declination (Dec or δ). Latitude-type coordinate used to indicate the position of an object in the sky. Declination is expressed in a similar manner as latitude. 0° declination represents the celestial equator (0° latitude is at the equator). The North Celestial Pole has declination +90°, the South Celestial Pole, −90°. Declination is used in conjunction with *Right Ascension* to specify coordinates for celestial objects. Notation example: +67° 15' 33". See *Right Ascension*.

Deep Sky Objects (DSO). Refers to *galaxies*, *nebulae*, *globular clusters* and *open clusters*. Although the term connotes objects that are distant and faint, some deep sky objects are "bright" and span a large area of the sky. The Andromeda galaxy, for example, spans more than six Moon diameters and can be seen with the naked eye. The stars (including double and binary stars) and Planets (including all other members of our solar system) are not considered deep sky objects.

Dobsonian. A nickname for a Newtonian reflector on a simple altazimuth mount, popularized by John Dobson in the 1970s.

Drift Alignment. A method used to polar align an equatorial-type mount to a celestial pole. This method produces the most accurate type of polar alignment, but it generally takes the most time to accomplish. The process requires looking through a reticle eyepiece (that has at least one inscribed straight line) at a star on the southern meridian and then one lower in the eastern sky (for those in the northern hemisphere). In each case, you watch the selected star, and make adjustments in either the mount's *altitude* or *azimuth* until the star stays put on an east-west rule. The longer these stars stay on the east-west rule of the reticle eyepiece, the better the polar alignment. Visit www.whatsouttonight.com for links to detailed instructions.

Emission Nebula. See *Nebulae*.

Equatorial Mount (German Equatorial Mount). A type of telescope mount that facilitates observing and photographing celestial objects because only one axis has to move/rotate in order to keep an object centered in the eyepiece. Equatorial mounts have two perpendicular axes. The polar axis (*Right Ascension*) points to a celestial pole; the other axis (*Declination*) is positioned at 90° to the polar axis. Until the mid-1970s, the majority of telescopes at professional observatories had equatorial mounts. Today, most professional telescopes have computer-controlled *altazimuth* mounts because they cost less than equatorial mounts. Amateurs must use some form of an equatorial mount to take pictures of deep sky objects that require extended exposures (minutes to hours), otherwise, stars will not be pinpoint but will become elongated (called field rotation). The German Equatorial Mount (GEM), pictured on pages 59 and 345, is a specific design that is often used by amateurs. Also see *Setting Circles*.

Equilateral Triangle. A triangle whose sides are equal in length, meaning that all three sides have the same length. See an example on page 76 and *Isosceles Triangle*.

f/ratio. See *Focal Ratio*.

Field of View. Expressed in degrees (*arc degrees*), field of view can be true or apparent. True field of view is the actual amount of sky that can be seen through a telescope or binoculars. For example, if you look through a telescope and you see the whole Full Moon, nothing more and nothing less, then the true field of view is ½° or 30 arc minutes (30'). The arc degree number that expresses true field of view represents the arc angle diameter of the circle seen through the eyepiece. Apparent field of view is a design attribute of an eyepiece. The greater an eyepiece's apparent field of view (usually ranges from about 30° to 84°), the larger is the true field of view. A difference in apparent field of view is like the difference between looking out small and large windows. See page 71 for examples on calculating true field of view.

Flamsteed, John (1646–1719). An English astronomer who was appointed the first Astronomer Royal of the Old Greenwich Observatory just outside London (where 0° longitude starts). With a high degree of precision, he measured the position of thousands of stars. Instead of Greek letters, he used numbers to identify the brightest stars in the *constellations*. See *Flamsteed Numbers* and *Bayer Letters*.

Flamsteed Numbers. Numbers assigned to the brighter stars, by *constellation*, in order of Right Ascension, to aid in identification. Flamsteed numbers designate more stars than the *Bayer Letters*. See more on page 80. Also see *Flamsteed, John*.

Focal Length. For refractor and Newtonian reflector telescopes, the distance from the primary mirror or objective lens to the point where light comes to a focus. Usually expressed in millimeters. The focal length for catadioptric telescopes, like Schmidt-Cassegrain Telescopes (SCTs), is calculated differently because the optics in these systems multiply the individual focal lengths. The focal lengths of eyepieces are much shorter than telescopes and are always expressed in millimeters, generally ranging from 2mm to 55mm.

Focal Ratio (f/ratio). The ratio of the focal length of a telescope to the diameter of the primary mirror or objective lens. A focal ratio is calculated by dividing the diameter of the primary or objective into the focal length of the telescope (the units of measurement must be the same for the focal length and primary/objective diameter). Examples: f/1.9, f/4, f/5.4, f/8, f/11, f/15.5

Galactic Cluster. A term for a relatively young *open cluster* found in the spiral arms of a *galaxy*. Spiral arms have the highest concentrations of stars and hydrogen gas.

Galaxies and Their Classification.
Galaxy. If you held the Universe in your hands, you would see billions of tiny fuzzy specks. Each speck would be a galaxy, a basic grouping representing a collection of billions to hundreds of billions of stars which are gravitationally bound together. Their shapes are generally circular, being either flattish like a dish or dimensional like a ball. All galaxies lie outside our Milky Way Galaxy, the majority at distances of millions to billions of light years away. It is estimated that there are 125,000,000,000 (125 billion) galaxies in the Universe. The sizes of galaxies range from about 500 to 500,000 light years in length or diameter. Galaxies are clumped into clusters (our *Local Group* is composed of about three dozen galaxies) which are part of larger superclusters that stretch through the Universe. Galaxies in a cluster are gravitationally bound to one another. As a whole, galaxies in the Universe are distributed along interconnected strands, creating a network that looks somewhat like a sponge, where the holes represent areas devoid of galaxies.

There are three types of galaxies, classified by their shape. **Elliptical** galaxies are by far the most common and account for about 90% of galaxies. The smallest and largest galaxies are elliptical. They resemble balls or elongated balls. The stars in these galaxies revolve around their nuclei every which way, much like a swarm of bees.

The most distinctive looking galaxies are called **spirals**, and have several curved arms radiating from bulged centers or nuclei. These larger galaxies (on average) are shaped like a dish and account for about 5% of galaxies. Their distinctiveness comes from the fact that they have bright nuclei and arms. About 20% of all spiral galaxies are called "barred" spirals because they have what looks like a straight

Hubble Tuning Fork
Shape Classification of Galaxies

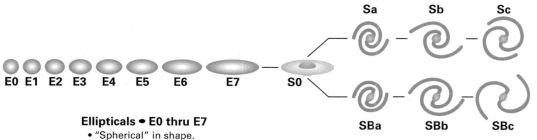

arm or "bar" passing through an elongated nucleus. The curved arms then radiate off the ends of the bar. Two percent of all spiral galaxies are Seyert galaxies, which have peculiar nuclei that are starlike — extremely small and bright. These galaxies have active nuclei from matter circling inward to a supermassive black hole.

Finally the remaining 5% of galaxies are called **irregulars** because they are irregularly shaped and have "mixed-up" insides. These galaxies appear to either result from the collision of two galaxies or represent a smaller galaxy being gravitationally contorted by a nearby larger galaxy.

It is believed that larger elliptical galaxies may be the result of the merger of two or more spiral galaxies, thus it is speculated that there may have been many more spiral galaxies when the Universe was younger. Data also suggest that most, if not all, galaxies have giant black holes at the center of their nuclei.

The Hubble "Tuning Fork," shown on the previous page, is the system used to classify the shapes of galaxies (there are subdivisions that are not noted here). Edwin Hubble developed this scheme after photographing and comparing numerous galaxies in the early 20th century. This classification has nothing to do with the evolution of galaxies, it's used only to specify their shapes.

Genitive or Latin Genitive (form of constellation names). Genitive refers to a grammatical case resulting in a spelling variation of a word that indicates or implies a relationship of possession. Genitives are not used in modern English. Where they exist, they are a holdover from Latin and other languages. Genitives of the *constellations* are still used today in conjunction with the Greek *Bayer Letters* and the *Flamsteed Numbers*. For example, the "alpha" star in Orion, best known as Betelgeuse, could also be referred to as α (alpha) Orionis, where Orionis is the genitive form of Orion, indicating possession of the alpha star. Betelgeuse can also be indicated as 58 Orionis when referring to it by its Flamsteed number. See pages 334–336 of this Appendix for a listing of the genitive forms of the constellations.

German Equatorial Mount. See *Equatorial Mount.*

Globular Cluster. This *deep sky object* is a closely packed group of 10,000 to a million or so stars resembling a ball (like cotton balls) that are gravitationally bound. Globular clusters are not *galaxies*, but are gravitationally attached to *galaxies*. It is estimated that there are about 200 of them surrounding our *Milky Way Galaxy*, almost all outside the plane. Some of the globular clusters that surround the Andromeda Galaxy are visible with amateur telescopes. The stars in globulars clusters are older, *Population II stars*, which means that they are poor in heavier elements. Globular clusters probably formed along with *galaxies*.

Globular clusters can be classified by how tightly their stars are concentrated or bunched together. One scheme, the Shapley-Sawyer, indicates this by using the Roman numerals I thru XII where I is the most compact (tight like a ball) and XII the least (more like a concentrated *open cluster*). In the catalogue section of this book, under the heading "Facts," I use this scheme to describe the compactness of globular clusters, but I replaced the Roman numerals with words like tight and loose.

GO TO (also written as GOTO). GO TO refers to a telescope's motorized mount that has capabilities for automatically moving (slewing) to a celestial object selected from a hand controller (the telescope itself is not GO TO, only the mount). The hand controller looks similar to a phone pad with a display screen that allows access to catalogues listing hundreds to thousands of objects, including the Planets.

Great Attractor. A distant and huge concentration of *galaxies* about 200 million light years away in the direction of the *constellations* Hydra and Centaurus which contains a total mass 100,000 times the mass of our *Milky Way Galaxy*. The gravity from this concentration influences our own *Local Group* of three dozen *galaxies*.

Great Circle. Any circle on the *celestial sphere* that divides it into equal halves. The *celestial equator* is a great circle. Any circle that passes through both celestial poles is a great circle. A great circle is the largest circle that can be drawn in the sky.

Great Wall. A huge concentration of *galaxies* 250 million light years away strung along a "line" 600 million *light years* in length. It is north of the *celestial equator* between RAs 8h and 17h. Galaxy clusters in Hercules and Coma Berenices are part of the wall (the Virgo Galaxy Cluster is not part of the Wall).

Greek Alphabet or Letters. See table on page 81 and the last page of this book.

Hubble Space Telescope. Launched into orbit 375 miles (600 km) above Earth in 1990, this 94-inch (2.4 meters) diameter telescope has revolutionized astronomy by providing the most detailed images of the Universe yet.

IC (Index Catalogue of Nebulae and Star Clusters). The IC catalogue is a huge addendum to the original *NGC* catalogue compiled by J. L. E. Dreyer and refers to two supplemental catalogues listing 5,386 *deep sky objects*. In Dreyer's IC catalogue, each object is identified with the letters IC followed by a number from 1 through 5,386. The first *Index Catalogue*, published in 1895, lists 1,529 objects and the *Second Index Catalogue*, published in 1908, lists 3,857 objects. The majority of IC objects are faint *galaxies*. See *NGC*.

Isosceles Triangle. A triangle that has two sides equal in length. See an example on page 76, and *Equilateral Triangle*.

Latin Genitive. See *Genitive*.

Light Year (ly). Unit of length in astronomy. One light year is the distance light travels in one year. Since light travels at the rate of 186,282 miles per second, it will travel approximately 6 trillion miles (5,880,000,000,000 miles) in one year's time. It takes light 1.3 seconds to travel the distance from the Earth to the Moon and 8.3 minutes to travel from the Sun to the Earth. Our solar system is about 11 light hours in diameter, the nearest star is about 4 light years away and our galaxy is about 80,000 light years in diameter.

Local Group. A group of about three dozen *galaxies* that includes our *Milky Way Galaxy* and the Andromeda Galaxy. The Local Group galaxies are gravitationally bound together. *Galaxies* are clumped together in clusters throughout the Universe.

Meridian. See *Celestial Meridian*.

Messier, Charles (1730–1817). A famous French comet hunter of the late 1700s who was interested in all aspects of astronomy. One of his many accomplishments was publishing a catalogue listing 103 of the brightest *deep sky objects* visible from the northern hemisphere.

Messier Object. Any of the 112 deep sky objects catalogued or known by Charles Messier in the late 1700s.

Meteor. The light trail in the sky created when a meteoroid enters Earth's atmosphere, commonly called a shooting star. These trails are most often generated from meteoroids the size of a grain of sand — often debris from the wake of comets that have passed close to the Sun.

Milky Way Band. An irregularly shaped, hazy, cloudy or milky band that circles the *celestial sphere*. This band is a permanent part of the sky and has an average width of five *arc degrees* (10 Moon diameters). The Milky Way Band is impossible to see in larger cities because of light and air pollution; however, it is prominent in dark skies. Its appearance is milky because it is composed of the faint glow from countless faraway stars — the bulk of the stars in our *galaxy*. With a telescope or binoculars, one can see that there are many more stars in the region of the Milky Way Band than in other areas of the sky. Our *galaxy* took on the name of this band after it was realized that the Milky Way Band is our *galaxy*.

Milky Way Galaxy. The name of the *galaxy* that our Sun resides in. Our *galaxy* is classified as an SBc spiral (see page 340). It contains about 100 billion stars and probably has a "visual" diameter closer to 80,000 ly than the 100,000 ly that I and others often use.

Mirror Reversed Image. An image that has left and right reversed but maintains the correct vertical orientation. This is the image seen in mirrors. Refractors and Schmidt-Cassegrain Telescopes that use 90° diagonals for comfortable visual observing have mirror-reversed imagery. The biggest drawback to this type of imagery is the difficulty it presents when trying to correlate stars with star charts printed with a "correct" orientation.

Moments of Clarity. An atmospheric phenomenon that is noticeable when visually observing through a telescope, especially when viewing the Planets, Moon and Sun. Moments of clarity are usually frequent but split-second occurrences when the sky is steady and details of a Planetary surface (or other celestial detail) can be glimpsed.

Nadir. The point directly below an observer. We each have our own nadir. Opposite is the *zenith*, the point directly over our head.

Nebula (plural: nebulae). The word nebula is a general term referring to gaseous hydrogen clouds that reside inside *galaxies*. For the most part, the only nebulae that can be observed are the closest — those in our *Milky Way Galaxy*.

There are three basic types of nebulae: galactic clouds, planetary nebulae and supernova remnants.

Galactic clouds, a term that I use, represent birthing places of stars. They can be lit or unlit. Those that are lit are bluish or reddish in color because they either reflect starlight or emit their own light. Reflection nebulae look bluish in color from reflecting the light of nearby stars. Emission nebulae look pinkish because their hydrogen is

The Horsehead Nebula (IC 434), located off the left belt star in Orion, is a dark nebula that is in front of an emission nebula.

N. A. SHARP/AURA/NOAO/NSF

stimulated to give off its own light from nearby, highly energetic starlight. Galactic nebulae are located mostly in the arms of our *Milky Way Galaxy*. Those nebulae that are unlit are often seen as silhouettes. A good example of a reflection nebula is the Trifid Nebula, M16, and one of an emission nebula is the Great Orion Nebula, M42. The best example of a dark nebula is the Horsehead Nebula pictured here, which requires about a 20-inch diameter scope to see visually.

Planetary nebulae (the name has nothing to do with the Planets, it's just an old name that stuck because many of these nebulae look like small round disks) represent the remains of outer atmospheres shed from large stars in their death throes. These nebulae are spherical, ringed or have diametrically opposed lobes depending on magnetic fields generated by the dying stars. The most well-known planetary nebula is the Ring Nebula, M57.

Supernova remnants are nebulae created from the explosions of very large stars at the end of their lives. The Crab Nebula, M1, is a good example. Supernova remnants are rare compared to the other nebulae.

NGC (New General Catalogue of Nebulae and Star Clusters). A listing of 7,840 deep sky objects published by J. L. E. Dreyer in 1888 that is still used today. The majority of NGC objects are *galaxies* and *open clusters*. Overall, NGC objects are much fainter than Messier objects but a "little" brighter than Dreyer's list of 5,386 *IC* objects (Index Catalogue of Nebulae and Star Clusters). See *IC*.

Open Cluster. A *deep sky object* that is a grouping of stars, anywhere from a dozen to several thousand, that were born together (from the same *nebula*) and reside in close proximity to one another. Several open clusters are visible to the naked eye of which the Pleiades, M45 in Taurus, is the best example.

Parallelogram. A four-sided figure whose opposite sides are equal in length and parallel to one another. Both pairs of opposite sides must be parallel for the figure to be considered a parallelogram. Squares and rectangles are parallelograms. See an example on page 76.

Parsec (abbreviation pc). A unit of distance in astronomy equal to 3.26 *light years*. This length is based on the Astronomical Unit (the average distance from the Earth to the Sun or 92,955,800 miles). An Astronomical Unit would have to be moved 3.26 *light years* in distance for it to extend for an arc angle of just 1 second in the sky (remember, there are 3,600 *arc seconds* in an *arc degree*). Mpc is the abbreviation for a million (1,000,000) parsecs.

Peculiar Galaxy. A *galaxy* that substantially departs from the norm with perhaps an unusual nucleus, arms or other peculiarity.

Planetary Nebula. See *Nebula*.

Planisphere. A circular star chart that is used to find the *constellations*. The word planisphere refers to a sphere of stars plotted on a flat surface or plane. Planispheres are handy charts for beginners and amateurs because unlike star charts in books, they can be set to show the stars visible for any hour and day of the year. One of the most popular planispheres is the "David H. Levy Guide to the Stars."

Plössl Eyepiece. A popular optical design for eyepieces. It utilizes four lens elements in two groups and provides a large apparent *field of view* (50° or greater).

Population I and II Stars. Population I stars, which include our Sun, are relatively young stars that contain a higher abundance of metals compared to the older Population II stars. The metals contained in Population I stars came from *supernova* explosions of very large Population II stars. Population I stars are

David Levy Planisphere

often found in the arms of spiral *galaxies* where there is a higher concentration of *nebulae*. *Globular clusters* are composed mainly of the older Population II stars.

Position Angle. The relative orientation of a celestial body in the sky measured counterclockwise from 0° to 360° where 0° points to celestial north. Often used to describe double star positions.

Resolved. For deep sky observing, the ability of a telescope to image distinct stars. Larger telescopes can image fainter stars than smaller telescopes. A cluster of faint stars in a small telescope appears as a faint patch of light (unresolved or not resolved) instead of many distinct stars.

Right Ascension (RA or α). West-to-east coordinate used in conjunction with *Declination* to determine the position of objects in the sky. Right Ascension is analogous to longitude but uses different increments. Its divisions are based on the 24 hours of a *sidereal day*. Each hour is further divided into 60 minutes and each minute into 60 seconds. Zero hours (0h) passes through the vernal equinox in Pisces. Examples of RA are 12h 23.7m and 1h 14m 23s.

Seeing. A measure of the steadiness of the atmosphere based on the scale 0 to 10 where 0 represents extreme turbulence and 10 a velvety smooth steady sky. I think it is easiest to gauge seeing by viewing a Planet "significantly" above the horizon. A seeing of 0 would indicate that no surface detail can be seen: the image appears like a blur and there are no *moments of clarity*. A seeing of 10 means that you can plainly view surface details without any blurring of features produced from atmospheric movement (rare). Poor seeing affects viewing of the Planets more than of *DSOs*. See *Transparency*.

Setting Circles. Circular, graduated scales that are sometimes attached to the *Right Ascension* and *Declination* axes of equatorial mounts to aid in locating celestial objects by using their RA and Dec coordinates. These were more popular before the advent of the GO TO mounts. See picture on next page.

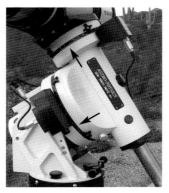

The two arrows point to the setting circles (black bands with markings) on this German equatorial mount. The upper setting circle is inscribed for Declination and the lower, larger setting circle has markings for Right Ascension.

Some fork-mounted SCTs also have setting circles but they are not functional unless the mount is titled forward on a "wedge," configuring it as an equatorial mount.

Seyfert Galaxy. See *Galaxies*.

Shooting Star. Common name for a *meteor*. See *Meteor*.

Sidereal Day. The amount of time it takes the Earth to make one complete rotation on its axis (23h 56m 4.1s), which is almost 4 minutes shorter than our normal clock time. Sidereal time is the pace at which the stars move across the sky and is measured as the elapsed time between two *meridian* passages of a star. Our normal clock time is based on the average of successive *meridian* passages of the Sun. Motorized telescopes move at the sidereal rate.

Star Cluster. A general term that refers to an *open cluster, galactic cluster* or *globular cluster*.

Supercluster. A group of gravitationally bound *galaxies* is called a galaxy cluster. There are also strings of galaxy clusters which form superclusters. Superclusters can stretch across a significant portion of the Universe.

Supernova (plural: Supernovae). An explosion of a massive star, at the end of its life, of such intensity that the light emitted outshines the star's *galaxy*. A supernova can remain brilliant for several weeks. They occur infrequently in our *galaxy*, so amateur and professional astronomers observe them more often in other *galaxies*. The last supernova visible in our *galaxy* was seen in the year 1604. Supernova explosions leave *nebulae* remnants. The best examples are the Crab Nebula, M1 and the larger Veil Nebula (NGC 6960/6992). See *Nebula*.

Transparency. A measure, given in magnitude, of the faintest star that can be seen with the naked eye at a particular time and location. The transparency in large cities is generally "poor" because various types of pollution prevent seeing fainter stars. Transparency is usually best in the country and on high mountains. Transparency, like *seeing*, can change in a "short" period of time depending on weather or other factors. See *Seeing*.

Twilight. The transition time between day and night either before sunrise or after sunset. There are three officially defined twilights — civil, nautical and astronomical. Each is based on the Sun's arc angle distance below the horizon.

Virgo Galaxy Cluster. A cluster of about 2,500 *galaxies* that resides in the direction of the *constellation* Virgo (as well as Coma Berenices) with distances averaging 56,000,000 light years. At its center is the large elliptical *galaxy* M87. This cluster is considered rich, crowded and centrally condensed, which probably makes for frequent galactic collisions.

Zenith. The highest point in the sky, directly overhead. Everyone has his or her own zenith (unless you are carrying someone on your shoulders, then you share a zenith). The opposite point is the *nadir*, which is the point beneath your feet.

Zodiac. Twelve *constellations* make up the zodiac. These *constellations* lie along a *great circle* in the sky called the ecliptic, the apparent path the Sun travels in the sky over the course of a year. The ecliptic is created from the Earth's yearly revolution around the Sun. There is no scientific significance to the *constellations* of the zodiac. The ecliptic is not indicated on the star charts in this book because it has no relevancy to finding or observing Messier *deep sky objects*.

List of Messier Objects

#	RA	Dec	Const.	Object	Mag.	Arc Size	Name
	YEAR 2000 COORDINATES						
M1	5h 34.5m	+22° 01'	Tau	Supernova Remnant	8	6' x 4'	Crab Nebula
M2	21h 33.5m	−0° 49'	Aqr	Globular Cluster	6.5	13'	
M3	13h 42.2m	+28° 23'	CVn	Globular Cluster	6.2	16'	
M4	16h 23.6m	−26° 32'	Sco	Globular Cluster	5.9	26'	Cat's Eye
M5	15h 18.6m	+2° 05'	Ser	Globular Cluster	5.7	17'	
M6	17h 40.1m	−32° 13'	Sco	Open Cluster	4.2	15'	Butterfly Cluster
M7	17h 53.9m	−34° 49'	Sco	Open Cluster	3.3	80'	
M8	18h 03.8m	−24° 23'	Sgr	Nebula	6	90' x 40'	Lagoon Nebula
M9	17h 19.2m	−18° 31'	Oph	Globular Cluster	7.7	9'	
M10	16h 57.1m	−4° 06'	Oph	Globular Cluster	6.6	15'	
M11	18h 51.1m	−6° 16'	Sct	Open Cluster	5.8	14'	Wild Duck Cluster
M12	16h 47.2m	−1° 57'	Oph	Globular Cluster	6.7	15'	
M13	16h 41.7m	+36° 28'	Her	Globular Cluster	5.8	17'	Great Hercules Cluster
M14	17h 37.6m	−3° 15'	Oph	Globular Cluster	7.6	12'	
M15	21h 30.0m	+12° 10'	Peg	Globular Cluster	6.2	12'	Great Pegasus Cluster
M16	18h 18.8m	−13° 47'	Ser	Nebula/Open Cluster	6	7'	Eagle Nebula
M17	18h 20.8m	−16° 11'	Sgr	Nebula/Open Cluster	7	46' x 37'	Omega Nebula
M18	18h 19.9m	−17° 08'	Sgr	Open Cluster	6.9	9'	Black Swan
M19	17h 02.6m	−26° 16'	Oph	Globular Cluster	6.8	14'	
M20	18h 02.6m	−23° 02'	Sgr	Nebula/Open Cluster	8	28' x 28'	Trifid Nebula
M21	18h 04.6m	−22° 30'	Sgr	Open Cluster	5.9	13'	
M22	18h 36.4m	−23° 54'	Sgr	Globular Cluster	5.1	24'	Great Sagittarius Cluster
M23	17h 56.8m	−19° 01'	Sgr	Open Cluster	5.5	27'	
M24	18h 16.9m	−18° 29'	Sgr	Thick Milky Way Patch	4	90' x 60'	
M25	18h 31.6m	−19° 15'	Sgr	Open Cluster	4.6	32'	
M26	18h 45.2m	−9° 24'	Sct	Open Cluster	8.0	15'	
M27	19h 59.6m	+22° 43'	Vul	Planetary Nebula	8	8' x 4'	Dumbbell Nebula
M28	18h 24.5m	−24° 52'	Sgr	Globular Cluster	6.8	11'	

#	RA	Dec	Const.	Object	Mag.	Arc Size	Name
	YEAR 2000 COORDINATES						
M29	20h 23.9m	+38° 32'	Cyg	Open Cluster	6.6	7'	
M30	21h 40.4m	−23° 11'	Cap	Globular Cluster	7.2	11'	
M31	0h 42.7m	+41° 16'	And	Spiral Galaxy	3.5	178' x 63'	Andromeda Galaxy
M32	0h 42.7m	+40° 52'	And	Elliptical Galaxy	8.2	8' x 6'	
M33	1h 33.9m	+30° 39'	Tri	Spiral Galaxy	5.7	62' x 39'	Pinwheel Galaxy
M34	2h 42.0m	+42° 47'	Per	Open Cluster	5.2	35'	
M35	6h 08.9m	+24° 20'	Gem	Open Cluster	5.1	28'	
M36	5h 36.1m	+34° 08'	Aur	Open Cluser	6.0	12'	
M37	5h 52.4m	+32° 33'	Aur	Open Cluster	5.6	24'	
M38	5h 28.7m	+35° 50'	Aur	Open Cluster	6.4	21'	
M39	21h 32.2m	+48° 26'	Cyg	Open Cluster	4.6	32'	
M40	12h 22.4m	+58° 05'	UMa	Double Star	9.6/10.1	1'	
M41	6h 46.0m	−20° 44'	CMa	Open Cluster	4.5	38'	Little Beehive
M42	5h 35.4m	−5° 27'	Ori	Nebula	4	66' x 60'	The Great Orion Nebula
M43	5h 35.6m	−5° 16'	Ori	Nebula	9	20' x 15'	
M44	8h 40.1m	+19° 59'	Cnc	Open Cluster	3.1	95'	Praesepe
M45	3h 47.0m	+24° 07'	Tau	Open Cluster	1.2	110'	Pleiades
M46	7h 41.8m	−14° 49'	Pup	Open Cluster	6.1	27'	
M47	7h 36.6m	−14° 30'	Pup	Open Cluster	4.4	30'	
M48	8h 13.8m	−5° 48'	Hya	Open Cluster	5.8	54'	
M49	12h 29.8m	+8° 00'	Vir	Elliptical Galaxy	8.4	9' x 7'	
M50	7h 02.8m	−8° 23'	Mon	Open Cluster	5.9	16'	
M51	13h 29.9m	+47° 12'	CVn	Spiral Galaxy	8.1	11' x 8'	Whirlpool Galaxy
M52	23h 24.2m	+61° 35'	Cas	Open Cluster	6.9	13'	The Scorpion
M53	13h 12.9m	+18° 10'	Com	Globular Cluster	7.6	13'	
M54	18h 55.1m	−30° 29'	Sgr	Globular Cluster	7.6	9'	
M55	19h 40.0m	−30° 58'	Sgr	Globular Cluster	7.0	19'	The Spectre
M56	19h 16.6m	+30° 11'	Lyr	Globular Cluster	8.3	7'	
M57	18h 53.6m	+33° 02'	Lyr	Planetary Nebula	9	1.3'	Ring Nebula
M58	12h 37.7m	+11° 49'	Vir	Spiral Galaxy	9.8	5' x 4'	
M59	12h 42.0m	+11° 39'	Vir	Elliptical Galaxy	9.8	5' x 3'	
M60	12h 43.7m	+11° 33'	Vir	Elliptical Galaxy	8.8	7' x 6'	

#	RA	Dec	Const.	Object	Mag.	Arc Size	Name
		YEAR 2000 COORDINATES					
M61	12h 21.9m	+4° 28'	Vir	Spiral Galaxy	9.7	6' x 5'	Swelling Spiral
M62	17h 01.2m	−30° 07'	Oph	Globular Cluster	6.5	14'	Flickering Globular
M63	13h 15.8m	+42° 02'	CVn	Spiral Galaxy	8.6	12' x 8'	Sunflower Galaxy
M64	12h 56.7m	+21° 41'	Com	Spiral Galaxy	8.5	9' x 5'	Black Eye Galaxy
M65	11h 18.9m	+13° 05'	Leo	Spiral Galaxy	9.3	10' x 3'	
M66	11h 20.2m	+12° 59'	Leo	Spiral Galaxy	9.0	9' x 4'	
M67	8h 51.4m	+11° 49'	Cnc	Open Cluster	6.9	30'	King Cobra
M68	12h 39.5m	−26° 45'	Hya	Globular Cluster	8.2	12'	
M69	18h 31.4m	−32° 21'	Sgr	Globular Cluster	7.6	7'	
M70	18h 43.2m	−32° 18'	Sgr	Globular Cluster	8.1	8'	
M71	19h 53.8m	+18° 47'	Sge	Globular Cluster	8.2	7'	
M72	20h 53.5m	−12° 32'	Aqr	Globular Cluster	9.3	6'	
M73	20h 58.9m	−12° 38'	Aqr	4-Star Asterism	10.5 (Brightest)	1'	
M74	1h 36.7m	+15° 47'	Psc	Spiral Galaxy	9.2	10' x 9'	The Phantom
M75	20h 06.1m	−21° 55'	Sgr	Globular Cluster	8.5	6'	
M76	1h 42.4m	+51° 34'	Per	Planetary Nebula	11	2' x 1'	Little Dumbbell
M77	2h 42.7m	−0° 01'	Cet	Spiral Galaxy	8.8	7' x 6'	
M78	5h 46.7m	+0° 03'	Ori	Nebula	8	8' x 6'	
M79	5h 24.5m	−24° 33'	Lep	Globular Cluster	7.7	9'	
M80	16h 17.0m	−22° 59'	Sco	Globular Cluster	7.3	9'	
M81	9h 55.6m	+69° 04'	UMa	Spiral Galaxy	6.8	26' x 14'	
M82	9h 55.8m	+69° 41'	UMa	Irregular Galaxy	8.4	11' x 5'	Cigar Galaxy
M83	13h 37.0m	−29° 52'	Hya	Spiral Galaxy	8	11' x 10'	
M84	12h 25.1m	+12° 53'	Vir	Elliptical Galaxy	9.3	5' x 4'	
M85	12h 25.4m	+18° 11'	Com	Elliptical Galaxy	9.2	7' x 5'	
M86	12h 26.2m	+12° 57'	Vir	Elliptical Galaxy	9.2	7' x 5'	
M87	12h 30.8m	+12° 24'	Vir	Elliptical Galaxy	8.6	7'	Virgo A
M88	12h 32.0m	+14° 25'	Com	Spiral Galaxy	9.5	7' x 4'	
M89	12h 35.7m	+12° 33'	Vir	Elliptical Galaxy	9.8	4'	
M90	12h 36.8m	+13° 10'	Vir	Spiral Galaxy	9.5	10' x 5'	
M91	12h 35.4m	+14° 30'	Com	Spiral Galaxy	10.2	5' x 4'	
M92	17h 17.1m	+43° 08'	Her	Globular Cluster	6.4	11'	

#	RA	Dec	Const.	Object	Mag.	Arc Size	Name
	YEAR 2000 COORDINATES						
M93	7h 44.6m	−23° 52'	Pup	Open Cluster	6	22'	
M94	12h 50.9m	+41° 07'	CVn	Spiral Galaxy	8.1	11' x 9'	Croc's Eye
M95	10h 44.0m	+11° 42'	Leo	Spiral Galaxy	9.7	7' x 5'	
M96	10h 46.8m	+11° 49'	Leo	Spiral Galaxy	9.2	7' x 5'	
M97	11h 14.8m	+55° 01'	UMa	Planetary Nebula	11	3'	Owl Nebula
M98	12h 13.8m	+14° 54'	Com	Spiral Galaxy	10.1	10' x 3'	
M99	12h 18.8m	+14° 25'	Com	Spiral Galaxy	9.8	5'	
M100	12h 22.9m	+15° 49'	Com	Spiral Galaxy	9.4	7' x 6'	The Mirror
M101	14h 03.2m	+54° 21'	UMa	Spiral Galaxy	7.7	27' x 26'	Pinwheel Galaxy
M102	15h 06.5m	+55° 46'	Dra	Elliptical Galaxy	9.9	6' x 3'	Méchain's Lost Galaxy
M103	1h 33.2m	+60° 42'	Cas	Open Cluster	7	6'	
M104	12h 40.0m	−11° 37'	Vir	Spiral Galaxy	8.3	9' x 4'	Sombrero Galaxy
M105	10h 47.8m	+12° 35'	Leo	Elliptical Galaxy	9.3	5' x 4'	
M106	12h 19.0m	+47° 18'	CVn	Spiral Galaxy	8.3	18' x 8'	
M107	16h 32.5m	−13° 03'	Oph	Globular Cluster	8.1	10'	
M108	11h 11.5m	+55° 40'	UMa	Spiral Galaxy	10.0	8' x 2'	
M109	11h 57.6m	+53° 23'	UMa	Spiral Galaxy	9.8	8' x 5'	
M110	0h 40.4m	+41° 41'	And	Elliptical Galaxy	8.0	17' x 10'	
M111	2h 19.0m	+57° 09'	Per	Open Cluster	4.5	30'	West Part of Double Cluster
M112	2h 22.4m	+57° 07'	Per	Open Cluster	4.5	30'	East Part of Double Cluster

PERMISSION IS GRANTED TO MAKE MULTIPLE COPIES OF PAGES 346 to 349 FOR INDIVIDUAL OR GROUP USE.

Messier Object Observing List

Object	Date Observed & Comments	Object	Date Observed & Comments
M1		M29	
M2		M30	
M3		M31	
M4		M32	
M5		M33	
M6		M34	
M7		M35	
M8		M36	
M9		M37	
M10		M38	
M11		M39	
M12		M40	
M13		M41	
M14		M42	
M15		M43	
M16		M44	
M17		M45	
M18		M46	
M19		M47	
M20		M48	
M21		M49	
M22		M50	
M23		M51	
M24		M52	
M25		M53	
M26		M54	
M27		M55	
M28		M56	

Object	Date Observed & Comments	Object	Date Observed & Comments
M57		M85	
M58		M86	
M59		M87	
M60		M88	
M61		M89	
M62		M90	
M63		M91	
M64		M92	
M65		M93	
M66		M94	
M67		M95	
M68		M96	
M69		M97	
M70		M98	
M71		M99	
M72		M100	
M73		M101	
M74		M102	
M75		M103	
M76		M104	
M77		M105	
M78		M106	
M79		M107	
M80		M108	
M81		M109	
M82		M110	
M83		M111	
M84		M112	

Greek Alphabet

	Lower Case & Variations*
Alpha	α
Beta	β
Gamma	γ, γ
Delta	δ
Epsilon	ε
Zeta	ζ, ζ
Eta	η
Theta	θ, ϑ
Iota	ι, ι
Kappa	κ
Lambda	λ
Mu	μ
Nu	ν
Xi	ξ, ξ
Omicron	ο
Pi	π, π
Rho	ρ
Sigma	σ
Tau	τ, τ
Upsilon	υ, υ
Phi	φ, φ
Chi	χ, X
Psi	ψ
Omega	ω

*Only the lowercase letters of the Greek alphabet are used for designating the brightest stars in each constellation.

Common Abbreviations

M (number)	Messier catalogue designation (numbered 1 to 112)
NGC (number)	New General Catalogue designation (numbered 1 to 7840)
IC (number)	Index Catalogue designation, an addendum of objects to the New General Catalogue (numbered 1 thru 5386)
RA or α	Right Ascension. Expressed using **h m s**. There are 24 hours of Right Ascension but these 24 hours are based on a sidereal day which is about 4 minutes shorter than clock time.
h m s	Hours, Minutes, Seconds. Ex: 11h 04m 59s
Dec or δ	Declination. Expressed using arc angles: ° ' "
° ' "	**Arc angle** Degrees, Minutes, Seconds. Ex: 36° 15' 23"
DSO	Deep Sky Object
ly or l.y.	Light Year (unit of length — almost 6,000,000,000,000 miles)
pc	Parsec (unit of length — 3.26 light years)
Mpc	Million Parsecs (3,260,000 light years)
AU	Astronomical Unit (unit of length — 92,955,800 miles)
mm	Millimeter (there are exactly 25.4 mm to an inch)
SCT	Schmidt-Cassegrain Telescope
APO	Apochromatic (best optics, generally refractors)
GEM	German Equatorial Mount
GO TO or GOTO	Computerized & Motorized Mount
altaz, alt/az, alt-az	Altazimuth (mount)
f/(number)	Focal Ratio (number = Focal Length ÷ Diameter). Ex: f/8